DATE DUE			

Household Pests

Household Pests

*A Guide to the Identification
and Control of Insect, Rodent,
Damp and Fungoid Problems
in the Home*

Peter L. G. Bateman
FRES

BLANDFORD PRESS
Poole   Dorset

First published 1979

© 1979 Blandford Press Ltd.
Link House, West Street
Poole, Dorset BH15 1LL

© Text Peter L. G. Bateman

British Library Cataloguing in Publication Data

Bateman, Peter L G
    Household pests.
    1. Household pests – Control – Great Britain
    I. Title
    628.9′6′0941   TX325

ISBN 0 7137 0915 4

Filmset and printed in Great Britain
by
BAS Printers Limited,
Over Wallop, Hampshire

Colour plates printed by Tonbridge Printers Ltd.

793657

For Rachel, Katie, Richard and Claire, the nicest household pests I have ever tried to control.

Contents

Acknowledgements

I acknowledge with thanks the help and encouragement given me by Angus Fraser McIntosh O.B.E., Past President of the British Pest Control Association and Founder President of the Confederation of European Pest Control Associations and I sincerely thank Dr Peter Cornwell, Dr Norman Hickin and my co-directors of Rentokil Ltd for permission to make use of some of the company's published material, facilities and illustrations. The photograph on page 27 is reproduced by permission of A. T. Paton, and those on pages 13 and 41 by courtesy of the Tunbridge Wells, Museum. The line drawings are by David Dowland and Joyce Smith.

I also thank June Halls at the Rentokil Advice Centre for showing me the need for such a book and to Stephen Dalton F.R.P.S., who has provided some of his superb photographs to help illustrate it.

I am also grateful to Robin Edwards F.R.E.S. for his advice over many years and to my wife Joan for understanding my neglect of the house and garden while I was writing this book.

I hope the work will be of some help not only to householders but to teachers and students of public health, catering, institutional management and home economics.

P.L.G.B.
October, 1978

Introduction

When we talk about our 'creature comforts' we are unlikely to be thinking about all the creatures that seek to share them with us in our homes. The warmth, food and shelter of a house can provide 'Home Sweet Home' for a surprising variety of wildlife—not all of it welcome.

No matter how well built a house may be and however well it is maintained, it is always liable to be invaded by unwanted insects or rodents. Many of these are harmless casual wanderers from the garden, some may be accidentally carried indoors, but others can become expensive lodgers. It is as well to be able to recognise those that can cause damage or spread contamination and it is essential to know how to get rid of them.

The human race creates its pests and the first field planted with one crop provided a bonanza for the insects for which it was a food plant. Volumes have been written on agricultural and garden pests as a consequence. The same principle applies to buildings. From the time the first man went to live in a cave, creatures have moved in with him. Some crawled, some flew, some went in with animal skins and tree branches. One or two specialised forms developed which literally came to live on us. Fleas, lice and bed bugs may truly be said to associate with man for what they can get out of him: others are pests because they eat our food or bring disease organisms into the house. Others again damage fabrics, furniture, even the structure of a building, and some we dislike having around because they are psychologically disturbing.

The parasites include lice and bed bugs; those which infest food include larder beetles, flour moths, cockroaches. There are those which damage textiles, clothing and soft furnishings, such as clothes moths and carpet beetles; and those such as mice, flies and wasps which invade

9

A Ratcatcher 1560.

premises at certain specific times of the year. Casual intruders such as maybugs, lacewings, earwigs, woodlice, clover mites and ground beetles may also enter accidentally. Woodworm, termites and rodents cause physical damage; flies, cockroaches and mice are a threat to health as they carry disease organisms; spiders, furniture mites, wasps or silverfish produce anxiety and distress. Feral pigeons may simply be a nuisance.

Improved standards of living and environmental hygiene have greatly reduced infestations by parasites and many flies, yet developments such as modern central heating, fitted carpets and the continued use of untreated softwood timbers tend to contribute to the growing problems caused by carpet beetles and woodworm. Tropical species like cockroaches now thrive in man-made moist warm climates under hotels, hospitals or in commercial kitchens in temperate zones. There are very many other examples of the way pests change in importance. The replacement of the horse by the motor car probably decimated the fly population. More recently the synthetic insecticides have added their contribution. The development of resistance to certain pest control chemicals, and restrictions on the use of some alternatives have, however, caused a great increase in mouse infestations and made the control of some insects more difficult.

Greater awareness of the need for hygiene in food manufacture and catering has made us less tolerant of the one or two flies or the occasional mouse dropping in industrial or commercial premises, yet our own homes often provide harbourages for potentially dangerous creatures.

Fear of insects in buildings is very real and very widespread, even if not always rational. In a survey of nearly 3,000 people in the USA to discover what people most fear, 'insects and bugs' ranked third equal with financial problems. More people were afraid of insects than of loneliness, dogs, cars, lifts, sickness, darkness—or death! In severe cases, entomophobia can become a disabling psychological problem and some sufferers fear entirely imaginary insects attacking their property and their persons.

There is no doubt that changes in the types of materials used in building construction, the design of buildings and the increased use of central heating are contributing to changes in the incidence of pests in buildings.

No longer do large areas of thatch and wattle and daub harbour resident populations of rodents and birds with their attendant insects.

Massive over-mature rafters and joists of hardwood which delighted the Death Watch Beetle have given way to small cross-section timbers of softwood, an ideal diet for the Common Furniture Beetle. Indeed, to have Death Watch today is practically a status symbol!

Far from 'building out' pests by sensible design at the drawing board stage, we now install boxed in pipe-runs, false ceilings, concealed ducts and service conduits which provide motorways for mice and highways for cockroaches, with access to all parts.

The artificial micro-climate created within central heated buildings, especially where catering is carried on, produces warm humid conditions in which pests of tropical origin such as cockroaches and Pharaohs ants can live regardless of the outside temperature.

The length of life cycle of all insects is very dependent upon temperature and humidity, more generations being produced in warm conditions, and insects such as cat fleas that were once a purely seasonal problem are now active all the year round indoors.

The greater use of synthetic fibres has reduced the risk of moth damage to clothes and soft furnishings but the increased acreage of wall to wall carpets has encouraged the 'woolly bears' picnic' by the hairy grubs of the carpet beetles.

Warfarin resistance—a more recent problem.

In recent years, the development of new rodenticides has become both desirable and necessary. In the UK, the Animals Cruel Poisons Act 1962 which banned phosphorus, red squill and arsenic for rodent control, led to the almost total reliance on warfarin and the consequent development of warfarin-resistant rats and particularly mice,—the so-called 'Super-mice'.

In Britain the same Act requires that rats and mice must not be killed by chemicals that cause them symptoms of pain or suffering, they must literally be 'killed with kindness'. The ideal mouse-killer must therefore be toxic to mice, but not to cats, dogs or children; be palatable to mice and so attractive to them that they eat it in preference to whatever else is around them in a supermarket, warehouse or kitchen larder; and must kill the mice humanely and without pain.

The chemicals used for rodent control are mostly slow-acting anti-coagulants such as warfarin, coumatetralyl, difenacoum or chloropha-cinone. The quick-acting alphachlorolose induces sleep and causes death by hypothermia but in very warm premises mice may recover and in such

An 18th-century wooden
mouse trap.

areas calciferol should be used.

Similarly, insecticides must be as far as possible specific to the target species; sufficiently long-lasting to deal with grubs that may hatch six weeks after a first treatment, and to give protection from re-infestation, but not so persistent that they contaminate the environment with undesirable residues.

Account must also be taken of insects developing immunity to certain insecticides and indeed to whole groups of related insecticides.

The chemical insecticide D.D.T. has long been superseded by other organo-chlorine insecticides such as dieldrin, chlordane and lindane (gamma HCH); organo-phosphorus insecticides of low mammalian toxity such as malathion, diazinon, dichlorvos, iodofenphos, fenitrothion; and carbamates such as carbaryl or propoxur.

The most promising new chemical insecticides are the synthetic, long-lasting pyrethrins. Fumigation of infested commodities or warehouses is now carried out with methyl bromide or Phostoxin aluminium phosphide tablets releasing phosphine gas.

Among the 'alternative pesticides', investigations have been made into the use of ultrasonics, chemo-sterilants and nuclear radiation, none of which has proved practicable. However, work on synthetic sex attractants, aggregation pheromones and juvenile hormones which modify the behaviour or development of insects may lead to more efficient and even safer techniques for the future protection of health, food and property.

PART I Background to Pests and Problems

Why Bother?

Pests are unacceptable in buildings for four very good reasons that can be put under the headings of Legislation, Contamination, Depreciation, Reputation.

Legislation

In the UK, under the Prevention of Damage by Pests Act 1949, owners or occupiers of premises are bound to keep them free from large numbers of rats or mice or to report serious infestations to the local authority.

For premises where food is manufactured, stored, packed, transported or sold the Act also covers infestation by mites and insects.

For food manufacturers and retailers the Food and Drugs Act 1955 makes it an offence to sell contaminated food and in catering establishments the Food Hygiene Regulations 1970 emphasise the importance of freedom from pests in areas where food is prepared or served. It is estimated that only about 10 percent of cases of foreign bodies in food are reported to local authorities, yet each year there are about 14,000 prosecutions under these Acts. Probably half the foreign bodies are contamination by insects or rodents. The passing of the 1976 Food and Drugs (Control of Food on Premises) Act gives local authorities the power to close dirty food shops and restaurants in 72 hours and evidence of pests is generally considered as evidence of unhygienic conditions.

Contamination

Flies, cockroaches, mice and other intruders carry disease bacteria on or within their bodies. These bacteria are capable of causing food poisoning and more serious illnesses as the carriers commute from filth to food. In Britain in 1976, there were 12,000 notified cases of food poisoning—and

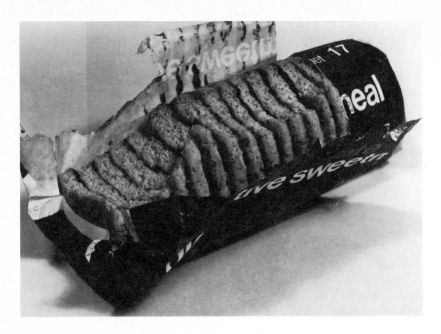

Biscuits that made a meal for a mouse.

many more go unreported.

Mice pollute far more food than they eat. Their droppings are fairly obvious, but they also have uncontrollable bladders. Insect bodies, moulds, rodent hairs can also contaminate our food or the environment of the home.

Depreciation

We cannot afford to let woodworm eat us out of house and home or allow the woolly bear grubs of the carpet beetle to make a meal of the Axminster. Textile pests and woodborers cause millions of pounds worth of damage each year.

In those countries where they occur, termites are estimated to cause £600 million ($1200M) of damage every year. Rodents damage pipes and

Rat damage to
packaging.

cable insulation by their gnawing. The myriads of beetles, weevils and warehouse moths destroy commodities by the ton or cause such deterioration that the produce is only fit for cattle.

It has been said that the Mediterranean Flour Moth destroyed more stored cereal in wartime Britain than was lost by enemy action and that rodents and insects today destroy more than 10 percent of the world's cereals *after* harvest,—enough to feed 260 million people if it could be saved.

Reputation

Having pests is bad public relations. People think twice about buying from a baker who was reported in the local paper as being fined because a mouse dropping was found in one of his rolls.

Responsible commercial organisations go to great lengths to protect their goodwill by ensuring that the risk of infestation is kept to the minimum.

One of Britain's largest multiple retailers has published the true costs of eleven instances when it stopped production by food suppliers because

The Mediterranean
Flour Moth.

Larvae of the
Mediterranean Flour
Moth.

mice were present, in order to maintain its good name. The costs in terms of lost production, recalled goods, cleaning and disinfesting, ranged from £20,000 ($40,000) to £125,000 ($250,000).

What Are The Major Pests?

In terms of requests for expert help in dealing with problems, the Common Furniture Beetle or Woodworm must take first place as the major household pest. Mice, flies and ants are probably next in importance.

In Britain, local authorities have a duty under the Prevention of Damage by Pests Act 'to take such steps as are necessary to secure so far as practicable that their district is kept free from rats and mice'. Many local authorities offer a service to deal with other pests and a survey was published in 1970 based on information from 671 local authorities on the incidence of pests they were asked to deal with. This indicated that rats, mice, flies, ants, feral pigeons, cockroaches and fleas were of the most importance to their Public Health departments. Textile pests and stored product insects are not normally considered of public health significance but have considerable economic significance. Store any commodity for long enough and something will come along and eat it.

A Pest Control Principle

A sound principle for keeping the risk of pest infestation to the minimum is contained in three words—Exclusion, Restriction, Destruction.

Exclusion

Keep them out, by maintaining the building in good order. Make sure there are no gaps around doors or windows, no broken drains or air bricks, no gaps where plumbing pipes pass through walls.

If flies or wasps are a serious problem, consider insect screens over windows for the summer. Keep dustbins well away from windows, and keep their lids on.

Check that you don't bring woodworm trouble into the house with secondhand furniture, tea chests or wickerwork.

Restriction

If pests do get in, make sure there are no areas where they can feed and breed. Do not accumulate paper, felt or fabric in the loft; do not leave 'empty' tins or bottles around. Dispose promptly of all food 'left overs' or spilt crumbs. Wrap all refuse and garbage before putting it in the dustbin. Clear out the larder regularly and do not let old stocks of ingredients accumulate. Scrupulous hygiene, waste disposal and stock control deprive pests of food and lodging. Spray the dustbin with insecticide regularly in Summer and take the vacuum cleaner into all the awkward corners and under heavy furniture occasionally. The vacuum cleaner is a valuable item of pest control equipment.

Here is a summary of points for the householder to check against, many of which are illustrated in the house opposite.

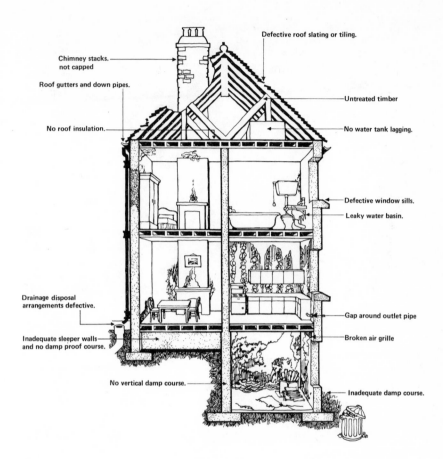

The house that has everything; some points for the householder to check.

Uncapped, disused chimney stacks enable birds and rain to enter.
Defective roof slating or tiling allows water, birds and insects to enter.
Blocked gutters and pipes may cause rot or allow insects to breed.
Broken lead flashing allows water to enter.
Untreated timber in roof is vulnerable to woodworm.
Lack of roof insulation and water tank lagging means heat is wasted.
Defective window sills encourage wet rot.

23

A pest control serviceman applies insecticide accurately and safely in a commercial kitchen.

A leaky water basin can cause rot and encourage mould and insects.
Wardrobes and carpets are vulnerable to moths and carpet beetles.
In the dining room scraps may attract ants, flies and mice.
Food in the kitchen should be protected to discourage flies or pantry pests.
Gaps round outlet pipes may let mice in.
Defective drainage arrangements provide entry for rats.
A broken air grille allows mice and rats to enter.
Stores of old papers etc. encourage mice and insects.
Inadequate damp coursing gives rise to the risk of rising damp and possibly dry rot in the flooring timbers.
Food scraps for birds and overflowing dustbins encourage rats and flies.

A regular annual check-up of the building structure will help reduce infestation.

Destruction

Despite research into physical and biological control methods, for most practical purposes in order to deal with household pests it is usually necessary to use some form of chemical. This may be a powder, an oil-based spray, an aerosol, a puffer pack, a bait or an insecticidal smoke. Fumigation may occasionally be necessary, but this can be undertaken only by specialist contractors.

Pest control chemicals are selected for their insecticidal or rodenticidal efficiency and their low toxicity to man and other animals. However, where food is prepared or stored take care not to spray utensils, working surfaces or the food itself.

Formulations sold for household purposes are safe when used as directed in accordance with the manufacturer's instructions. In Britain, reputable brands will have been cleared by the Ministry of Agriculture's Pesticides Safety Precautions Scheme set up in 1955.

The best way to deal with any insect depends upon many factors: the species of insect and its life cycle; where it came from and how it gets about; whether there are any circumstances such as the presence of exposed food, pets or children which might limit the use of certain insecticides. All these affect the decision on how to control it. The type of surface to which an insecticide is to be applied will also influence the choice of preparation, and it is necessary to know whether a quick immediate remedy will be sufficient or whether long term protection against re-infestation is also needed.

Flying insects such as house flies, mosquitoes or wasps need a quick-acting insecticide to give immediate clearance. Pyrethrins are present in many household insecticidal aerosols but until recently have had very little lasting effect. A new development, the synthetic pyrethroids, show far greater persistence and are likely to give prolonged freedom from flies.

Generally, especially where crawling insects or the grubs of flying insects are a problem, a fairly long-lasting insecticide is needed and it should be capable of remaining effective for about six weeks to deal with any pests emerging from eggs or pupae after the first application.

The constant risk of re-infestation from surrounding premises or suppliers means that for many problems the best remedy is a well trained pest controller visiting regularly.

Pest Control Precautions

'Take heede how thou laiest the bane for the rats,
Lest thou poison thy servant, they selfe and thy brats'.
Thomas Tusser, 1580
Five Hundred Pointes of Good Husbandrie.

1. Before using any insecticide or rodenticide in the home, read the instructions on the label and follow them exactly when carrying out treatment.

2. Store all pest control chemicals where children *cannot possibly* reach them.

3. Avoid transferring pesticides from their original containers if possible, but if it is done, return any unused residues to the proper container and do not leave any in unlabelled jars or tins. Before using an aerosol, make sure the hole on the release button is pointing in the right direction.

4. Liquid pesticides or solutions should never be placed in cups or glasses or bottles which might be reached by children. Any splashes of such liquid on to skin should be washed off, and after pesticides have been used hands should be washed before food is eaten.

5. Avoid inhaling vapour or fumes from sprays and do not allow pets or domestic animals to have access to baits, liquids or powders.

6. Many insecticides are carried in inflammable solvents, so do not spray near naked flames and do not smoke while applying them. This applies particularly to wood preservatives. Open windows and doors to improve ventilation.

7. Residues should be carefully disposed of by washing them down an outside drain with plenty of water. In the case of aerosols, *never* throw them on a fire or puncture them, but place the empty cans in a dustbin.

8. Burn or deeply bury the bodies of rats or mice killed by rodenticides.

9. Never apply insecticidal chemicals directly to the fur of cats or dogs unless they are specified for the purpose. Proprietary brands of flea powder or insecticide specially formulated for veterinary use are available.

Applying fly killer to surfaces where flies may land.

When Extra Help is Needed

The range of domestic pesticides available to the householder is necessarily limited and in the case of large infestations of any pests, or where insects seem difficult to kill, it is often wise to call on the services of a professional pest control servicing company. Such a company will have a wider range of materials than is available to the public, together with the special equipment and trained staff to apply them correctly and safely at the right dosage. The British Pest Control Association has a Code of Practice observed by its member companies and publishes a directory of members. Local Authority Environmental Health Departments will deal with rodents and some will cover certain insect infestations such as ants or wasp nests, fleas or bed bugs and pigeons.

Pest Control and The Law

'The health and safety of the people is the highest law'
Cicero *The Twelve Laws of Rome*

The Pesticides Safety Precautions Scheme

The control of pesticides in the UK is at the present time affected only to a limited extent by legislation. It depends mainly on the Pesticides Safety Precautions Scheme (PSPS).

The PSPS is voluntary and was developed and agreed between Government Departments and industry as represented by the British Pest Control Association and the British Agrochemicals Association. This Scheme came into formal effect in 1955.

The purpose of the Scheme is to safeguard human beings (whether they be users, consumers of treated produce, or other members of the public), livestock, domestic animals and wildlife, against risks from pesticides.

Manufacturers, formulators, and importers, voluntarily notify the Ministry of Agriculture, providing full toxicological details of any new pesticide or a proposed new use of an existing pesticide before introduction. They also agree not to introduce such materials until a decision has been reached on the appropriate precautionary measures and such precautions are included on the label. Companies complying with the scheme agree to withdraw a product from the market if recommended to do so by the Ministry of Agriculture on the advice of the Advisory Committee on Pesticides and other Toxic Chemicals.

It is a condition of membership of the British Pest Control Association that all members are bound to take part in the PSPS.

Thus, although this Scheme is voluntary, it works in practice as if it were mandatory and membership of the BPCA is a safeguard for the public and the hallmark of a responsible pest control company.

Rats are killed in sewers by placing poisoned bait on the ledge inside the sewer along which they run. Here, a Rentokil pest control team treat sewers for Westminster City council in London.

This Scheme applies to all active ingredients formulated as pesticides, that is insecticides, fungicides, herbicides, rodenticides or similar substances for use in agriculture, horticulture, forestry, home gardening, and food storage practice which includes warehouses, food factories, ships' holds, wholesale and retail food stores, and in larders and kitchens in institutions, hospitals etc. and the home.

Successive Ministers of Agriculture have praised the operation of the Scheme, and because of the whole-hearted co-operation of the pesticides industry and the pest control companies, the record of pesticide usage in the UK is second to none in terms of safety to all likely to be affected directly or indirectly.

The legislation which exists in the form of a number of Acts (and regulations made for their implementation) is set out below. Certain of these Acts are intended to protect the user and third parties, while others impose restrictions on the sale and supply of pesticides, in so far as the latter are classified as poisons. Specific legislation exists to protect the consumer against undesirable foodstuffs, to protect animals and birds and to prevent water pollution.

The relevant Acts include the following:

1. The Agriculture (Poisonous Substances) Act 1952 and Regulations.
2. The Agriculture (Safety, Health and Welfare Provisions) Act 1956 and Regulations.
3. The Hydrogen Cyanide (Fumigation) Act 1937 and Regulations.
4. The Farm and Garden Chemicals Act 1967 and Regulations.

Other laws relate to the protection of public health, food and property from the effects of pest infestation, the principle ones being:

1. The Prevention of Damage by Pests Act 1949
2. The Food & Drugs Act 1955
3. The Food Hygiene (General) Regulations 1970
4. The Public Health Act 1961
5. The Pharmacy and Poisons Act 1933 and the Poisons Rules.
6. The Food and Drugs Act 1955 and Regulations
7. The Protection of Animals Acts 1911 to 1927
8. The Protection of Birds Acts 1954 to 1967
9. The Health and Safety at Work Etc. Act 1974
10. The Food and Drugs (Control of Food Premises) Act 1976

Crown Exemption

State Hospitals and Schools were not, in the past, subject to existing legislation concerning hygiene and working conditions. They are now expected to conform to the requirements of the Health and Safety at Work Etc. Act, although they will be immune from the ultimate sanction of prosecution, as one Government department does not prosecute another. Virtually all other employers are bound by the Act to provide a health and safety policy which may well need to include protection from hazards arising from infestations.

The Health and Safety at Work Act is an enabling Act with powers to

The law-breaker. A House Mouse in a grocer's shop.

make it relatively easy for new regulations to be introduced by the Secretary of State and for repeals and modifications of existing legislation to be made. It also includes powers to modify the Building Regulations with regard to certain aspects including fire and public health. The Type Relaxation enabling the insulation of cavity walls to be carried out with appropriately certified materials was made under this Act.

These modifications and Codes of Practices will take some time to draw up and it is worth stating in some detail the provisions of some of the separate existing legislation that already affects pest control.

The Prevention of Damage by Pests Act 1949

This Act is concerned not with protecting health, but with preventing loss or damage by pests.

In this Act, paragraph 13 states that *'every person whose business consists of, or includes, the manufacture, storage, transport or sale of food, shall give to the Minister forthwith notice in writing if it comes to his knowledge that any infestation is present: (a) in any premises or vehicle, or any equipment belonging to any premises or vehicle, used or likely to be used in the course of that business for the manufacture, storage, transport or sale of food; (b) in any food manufactured, stored, transported or sold in the course of that business, or in any foods for the time being in his possession which are in contact or likely to come into contact with food so manufactured, stored, transported or sold.*

Directions may be made under the Act prohibiting or restricting the use for the manufacture, storage, transport or sale of food of any premises, vehicle or equipment which is or is likely to become infested. Directions on treatment may also be made to prevent or remedy infestation.

In a later section, paragraph 28 defines 'infestation' as 'the presence of rats, mice, insects or mites in numbers or under conditions which involve an immediate or potential risk of substantial loss of, or damage to, food . . .'. The Minister in this case is represented by the Local Authority. Local Authorities themselves have a duty under the Act to take such steps as are practicable to ensure that their district is kept free from rats and mice.

The Food and Drugs Act 1955 and Food Hygiene Regulations 1970

The Food Hygiene Regulations 1970 çover the prevention of contamination of food and lay down, in general terms, that premises where food is manufactured, prepared, packed or served shall be clean so as not to support any infestation. These Regulations are made under the Food and Drugs Act 1955, which deals with foreign matter in food which could render it unfit for consumption. They spell out that as a General Requirement (Part II, 6) 'No food business shall be carried on at any insanitary premises or place, the condition, situation or construction of which is such that food is exposed to the risk of contamination'.

Part IV, 25 states 'The walls, floors, doors, windows, ceiling, woodwork and all other parts of the structure of every food room shall be kept clean and in such good order, repair and condition as to enable them to be effectively cleaned and to prevent so far as is reasonably practicable, any risk of infestation by rats, mice or insects'.

The Health and Safety at Work and Control of Food Premises Acts

The Health and Safety at Work Etc. Act with its provision for wider powers and the immediate introduction of Improvement Notices and Prohibition orders already gives more powerful enforcement of the law and is obviously intended to involve management and staff very much more in this subject, of which pest control is an important part.

Until 1976, an Environmental Health Inspector often had to await the outcome of prolonged Court Proceedings before effective action could be taken against an offending food manufacturer, retailer or caterer.

Some authorities, however, had passed their own by-laws giving power to inspectors to close such premises pending Court proceedings. These included London (under the GLC General Powers Act), Manchester, Coventry, Derby and Gwent.

Under the Food and Drugs (Control of Food Premises) Act 1976, an environmental health officer can apply to a Court for an emergency closure order to take effect in 72 hours and there is a £400 ($800) fine for an owner who defies such an order. All the other penalties for breaches of the Food and Drugs Act 1955 stand. The Environmental Health officer must tell the Court specifically what defects are causing a real threat to health and must specify what must be done to make the premises safe.

After putting right any health risk, such as the presence of cockroaches, mice or flies, the restauranteur or shop owner may apply to the Court for the order to be lifted. And there are provisions for proper compensation if the owner is acquitted at the main Court hearing. Premises will not be allowed to reopen until the local health authority issues a certificate stating that the reasons for closure have been eliminated.

The Act relates to 'open food stored, sold or offered or exposed for sale' and dangers to health for carrying on a food business include 'insanitary or defective structure or fittings or fixtures or equipment or the infestation of vermin . . .'.

Guarantees

Some companies offer long term guarantees or warranties following their treatment against certain categories of pest such as woodworm, dry rot and termites.

These should be in addition to a client's Common Law rights and be assignable to the new owner when a treated property changes hands. They are useful documentary evidence of treatment so long as the company remains in business to back its guaranteed workmanship.

Some guarantees are now for a period of 30 years so it is sensible to choose a company of such a standing that it is likely still to be around in 30 years time.

The Basic Insect

This is a layman's book from which most technical entomological jargon has been eliminated. There are, however, certain terms common to the structure and biology of insects that must inevitably occur so an elementary guide to those terms is necessary.

Insects do not have an internal skeleton. Instead they have a tough outer skin, the cuticle, forming an external skeleton or integument of three layers. So that they can still move, their bodies consist of jointed

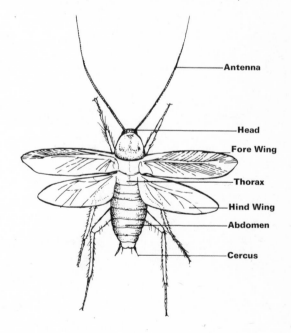

Main parts of an insect.

35

segments which are themselves divided into sections. The three main divisions of the adult insect are Head, Thorax and Abdomen.

The relative size and degree of movement between the basic sections differs considerably from one group of insect to another. Ants and bees for instance have a long narrow 'waist' between thorax and abdomen while other insects are equally wide along their length and have very limited movement between their sections.

The most primitive insects and some of the most specialised have a very simple life cycle, the 'nymphs' which hatch from their eggs resembling the adult in almost every respect except sexual maturity and size. These nymphs simply grow by a series of moults. This type of insect is said to be hemimetabolous. The majority of insects, however, undergo a metamorphosis from egg to larva (the 'grub' or 'caterpillar') to pupa (the 'chrysalis' or 'cocoon') before emerging in the final adult form, the imago.

These insects which change their forms so completely are called holometabolous. Growth only occurs in the larval stage and the grubs shed their skins as they grow. This is called ecdysis.

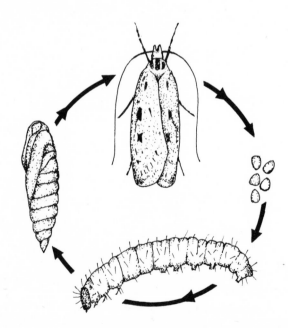

Life cycle of a House
Moth. (After Hickin)

Generally, the more primitive insects are wingless, but others such as fleas for whom wings would prove an encumbrance for the life style they have evolved, have lost the power of flight and retain only vestigial wings. Insects in the order Diptera (to which all the flies and gnats belong) have only one pair of wings, usually transparent. Others have two pairs of wings which may be clear, as in the dragonflies, or covered with coloured scales, as in the butterflies and moths. The beetles have a pair of hard wing cases covering a single pair of membranous wings. Insects 'breathe' through a series of small openings called spiracles along the sides of their bodies, and through a series of air-tubes called tracheal tubes. This system is not as efficient as a pair of lungs for getting oxygen to the various parts of the body and this is one reason why insects are small creatures.

The head of an adult insect contains the mouthparts which vary according to the feeding habits (and some adult insects do not feed at all), the eyes which in adult form usually consist of a large number of separate facets, and the antennae (the 'feelers') which are special sensory organs connected with taste, smell and touch.

The thorax carries the wings where these are present, and the characteristic three pairs of legs. Each leg has five distinct sections.

The abdomen normally consists of a series of round segments the last few of which carry the reproductive apparatus which is often diagnostic in identifying closely similar species.

Household Pests Past and Present

No longer do we build granaries on mushroom-shaped staddle stones to keep out rats. Even the ship's cat has been pensioned off. 'Bed bug destroyer to the Royal Household' is a title which, like 'The Imperial Entomologist', has passed away, overtaken by the Rodent Operative and the Pest Control Surveyor.

In the USA, the Exterminator now calls once a month, not to demand protection money but to see if you are harbouring cockroaches or suffering from Big Headed Ants or Depressed Flour Beetles.

The earliest forms of pest control were devised to limit the spread of epidemic diseases. The tolls of bubonic plague (The Black Death), typhus and other pest borne ills were recorded. Venice set up a Sanitary Council in 1348 and established a quarantine station in 1432. The word quarantine indicates the 40 day period of isolation imposed on ships, passengers and goods believed to be infected with one of the major diseases, principally plague or yellow fever and later cholera.

Drake found cockroaches aboard the 'Philip' and in the reign of Elizabeth I, a bounty was paid for dead rats. It is recorded that Casanova, returning from Corfu to Venice in 1745, was detained for 42 days in a quarantined ship.

Commercial interests became intolerant of the losses and delays imposed by misuse of quarantine regulations and in 1851 the first International Sanitary Conference was held by the main maritime nations. This Conference drew up regulations placing emphasis on ships arriving with a clean bill of health, but provided a ship came from a plague-free port and there had been no illness on board, passengers, crew and cargo could land without restriction. Detailed regulations against rats were introduced in 1903 after the connection between rat fleas and plague had been conclusively demonstrated in 1879 by Simond and the 1926 International Sanitary Convention laid the basis for the present day marine health regulations which are reviewed by the World Health Organisation (WHO). The Danes introduced a Rat Law in 1907 to encourage rat destruction on land, setting the scene for a whole new science.

Life and death for pests in the past was certainly colourful and spectacular if somewhat hazardous by present day standards. Trapping was one of the earliest forms of rodent control, but poisons appear earlier than we might think. Socrates mixed copper arsenate with bran to kill cockroaches, and Chaucer's Pardoner 'bought poyson his ratouns to quell'. Herbs also featured as the following extracts taken from the *Vermin Killer*, published in 1680, reveal, 'being a very necessary family book containing exact rules and directions for artificial destroying of vermin'.

To Kill Rats and Mice
Take Hellebore leaves mix with wheat flour making into a stiff paste with live honey and lay it into holes where rats and mice come and when they eat it it's pleasant Death. (Approved Paxamus)

A collection of ancient rat traps and mouse traps.

A modern galvanised mouse trap.

To drive away rats and mice from a house or other place, take the herb wild marjoram, called in latin origanum, and burn it in all the rooms of the house or place where you would be rid of mice and they will immediately depart and come no more for as long as the scent lasteth. (Approved Paxamus)

To Catch Moles
If you desire to catch moles, lay before the mole-holes a head of garlick or onion and they will immediately forsake their holes and may be taken by a dog. (Approved Albertus).

To Kill Snakes and Adders
Take a large radish and strike the snake or adder with it and one blow will kill them.

To Kill Pismires (Ants)
Take the roots of wild cucumbers and set them on fire where the pismires are and

Victorian glass fly traps.

the smoke will kill them. (Approved Agrippa)

To Kill Bugs
Take a convenient quantity of fresh tar, mix it with the juice of wild cucumber, let it stand a day or two, stirring it four or five times a day, then anoint the bedstead with it and all the bugs will die. (Approved Paxamus)

To Kill Fleas
Take wormwood and the roots of wild cucumber and boil them in pickle and sprinkle the room and it will certainly kill the fleas. (Approved Agrippa)

To Kill Flies
Take wild hellebore and steep in sweet milk, mix with it orpiment and sprinkle the room and places where flies come and they will die.
To gather the flies together take a deep earthen pot and lay in it beaten coriander and all the flies in the house shall be gathered together.

To Rid all sorts of Birds, Fowls, etc.
Take such seeds as the fowls or birds are wont to feed on and lay it in a soaking of Mother of Wine mixed with the juice of cicute and when it is well soaked through, throw it in the places where the birds or fowls feed and they will be

pleasantly drunk and lose their sence and you may take them with your own hand (Approved Albertus)

The same book states that troublesome Weasels 'can be frightened away by the smell of a burnt cat'. (So, presumably, could troublesome cats!) In 1730, John Southall of Southwark published *A Treatise on Buggs* and in 1775 Andrew Cooke of Holborn was claiming he had 'cured 16,000 beds with great applause'.

Today's pest controller is far removed from the Rat Catcher to the Princess Amelia, who in 1796 wrote *A Universal Directory For The Taking of Rats and all other Four Footed and Winged Vermin.* Nor can his white overalls match the resplendent red sash which once caused the Duke of Wellington to be told he looked like a rat catcher, the sash being a mark of office worn by John Black, Ratcatcher to Queen Victoria.

The age of psychedelic experiences and psychological disturbances has, however, produced its own complications for the men who protect food and property from insects and rodents and keep the pigeons off Parliament with strips of plastic jelly. To come more up to date, a survey carried out in 1976 by Environmental Health Departments for the Ministry of Agriculture, Fisheries and Food in Britain and covering 16,214 random inspections found 4·9 percent of all premises infested by mice and 4·8 percent by rats. In Metropolitan districts the figure for mice was 6·6 percent, but for rats only 3·7 percent, whereas in Rural areas the figures were 6·3 percent and 9 percent, respectively.

Apart from fairly crude inorganic chemicals for insects and rodents, a form of germ warfare was used to kill rats and mice with the Ratin bacillus discovered in 1902 and employed until warfarin took over in the 1950's.

In the Pied Piper town of Hamelin, the flute has been replaced by plastic bait depositor tubes to keep the city clear of rats.

Pest Control in the Space Age has had to prevent potential 'ratstranauts' stowing away on satellite launchers, repel cockroaches that board North Sea Oil drilling rigs and stop firebrats taking over a computer. The cloth cap of the ratcatcher has given way to the white coat of the Field Biologist.

Behind the field staff are the pest control laboratories where furniture beetle grubs are studied under X-rays, carpet beetles cut a rug in the interest of science, and intensive cultivation of moulds, the rearing of termites, the proliferation of cockroaches and grain weevils are all encouraged and cigarette beetles are fed on dog biscuits.

Words like 'specificity' are popular in pesticide research today, the aim being to control the target pest without harm or risk to anything else. Yet 'resistant' strains of insects, rats and mice are fighting back and the pests that trouble man are changing both genetically and in their occurrence.

No longer are we much troubled by the creature that once caused a Masonic Lodge to be known as 'the Fleamasons', but we still provide fodder for woodworm, and if further sign of an affluent society were needed, it can be found in the fact that the latest pest is—Mink.

PART II Insect and Rodent Pests, Rots and Moulds

Rodents

'And all through the house
Not a creature was stirring,
Not even a mouse'.   Clement Moore

Rats

The rat has plagued man for thousands of years, literally plagued him —
for the rat flea was responsible for the Black Death. The ratborne bubonic
plague still takes toll of human life in Asia. There are two common
species of rat, the Common or Brown Rat and the so-called Black or Ship
Rat. The Brown Rat is sometimes also called the sewer, Norway or water
rat.

Rodent droppings. *Left:*
Black Rat; *centre:* House
Mouse; *right:* Brown
Rat.

Identification

The Brown Rat *Rattus norvegicus* is the larger, often weighing 500g (over 1lb) and measuring about 23cm (about 9in) without the tail. It has a blunt muzzle, small hair-covered ears and a tail that is shorter than its body length. The Black Rat *Rattus rattus* weighs 225g (about ½lb) and is 20cm (about 8in) long excluding the tail, with a pointed muzzle, large almost hairless ears, a more slender body and a long thin tail that is longer than its body.

The Brown Rat is generally the commoner species and stays near ground level. The Black Rat still occurs in seaport towns and is a more agile climber, often entering the upper floor of buildings. It is the common species in parts of South Africa and the Far East. It is possible to

The Black Rat is an agile climber.

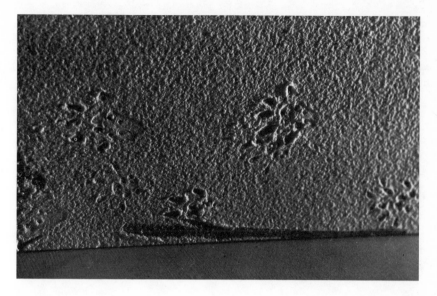

Rat footprints.

identify the species present from the different shaped droppings and the foot prints in dust (the Brown Rat is flat footed, the Black Rat runs on its toes). The droppings of the Black Rat are slightly curved, those of the Brown Rat, oval or spindle shaped. In towns, Brown Rats often live in sewers but in the countryside there is a constant background population in fields and hedges.

Life Cycle

Both species breed rapidly and become sexually mature in about three months. Each female may produce from three to twelve litters of between six and eight young in a year.

Brown Rats will burrow underground or into suitably soft material to make a nest. Refuse tips, loose soil under sheds, earth banks are likely sites and chewed paper, straw or insulation material may be incorporated as nest material.

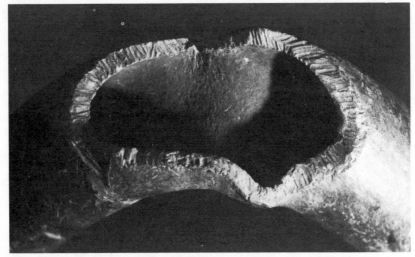

Lead water pipe gnawed through by a rat.

Rat damage to a loaf.

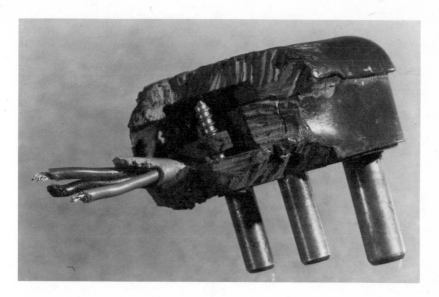

Rat damage to an electric plug.

The young are born blind, helpless and naked and depend on their mother for food for about three weeks before they take solid food.

Damage Caused

Rats, like mice, need to gnaw to keep their constantly growing incisor teeth worn down and they damage woodwork, plastic and lead pipes and will strip insulation from electrical cables by their gnawing. They are capable of spreading many diseases from their filthy surroundings in sewer or refuse tip and can transmit Salmonella food poisoning organisms, Weil's disease or leptospiral jaundice from which about ten people and a number of dogs die each year in the UK, Murine typhus, Rat Bite fever, Trichinosis and other diseases. They are probable carriers of foot and mouth disease on farms. In S.E. Asia and S. America their fleas still carry bubonic plague. They contaminate more food than they consume and their urine can pollute stagnant water with Weil's disease

Brown rats feeding on spilled grain.

organisms. Signs of their presence include dark droppings about 12mm ($\frac{1}{2}$in) long and black 'body smears' where they rub against a structure.

Rats will horde food for future consumption and numerous cases of 'theft' have been found to be the work of rats.

Rats feed mostly at night and an average rat will eat 57g (2oz) of food a day.

Creatures of habit, rats leave regular 'runs' to and from feeding areas. They can be a menace to poultry, eating eggs, chicks and feedingstuffs and will also eat almost any human or animal food.

Control Measures

Buildings must be rat-proofed by blocking all possible entry holes.

Workers in known rat infested areas of farms or in sewers should wear gloves and boots as a precaution.

As far as possible, eliminate harbourages such as gaps under sheds, loose piles of wood or neglected weed patches. Leaving scraps of food out of doors will encourage rats. If you feed garden birds, use a bird table or feeder basket. Where rats are found, one of the proprietary ready-mixed warfarin or calciferol or difenacoum baits should be used. These can be obtained as handy sachets which can be placed near signs of infestations. Keep replacing the baits until no more are taken. The warfarin at 0.05 percent concentration is evenly mixed through an oatmeal bait. For outdoor locations, sheds or damp buildings, Biotrol warfarin-impregnated wheat is a useful alternative and will not readily go mouldy. Calciferol whole grain baits contain 0.09 percent of the active ingredient. Rats avoid new objects so will take time to get used to break-back traps or baits.

For serious or persistent rat infestations, call a pest control servicing company or your local Public Health Department. They will have at their disposal other forms of rodenticides which are not available to the householder. In Britain, a free rodent control service for householders is offered by most local authorities but it is usual to charge for treating commercial premises.

Mice

The House Mouse, and sometimes the Long-tailed Field Mouse, seek the warmth and shelter of buildings for nesting sites and food. Traces of the small, grey House Mice first noticed by the householder are the presence of dark-coloured droppings or damage to stored foods, packaging or woodwork. Mice are now one of the most frequently encountered domestic pests.

Identification

The typical House Mouse *Mus musculus* is a small mammal, grey in colour, with prominent ears, whiskers and a long virtually hairless tail. The senses of hearing and smell are acute and the poor eyesight is compensated by the sensitive whiskers with which mice feel nearby objects.

Close-up of House Mouse.

Life Cycle

Mice are born in a nest of chewed paper, packaging or similar material. Mice become sexually mature in eight to ten weeks and a female may produce ten litters of up to sixteen young each. Born almost hairless and helpless, the young grow rapidly and soon after weaning leave the nest to seek food and harbourage elsewhere. They climb well and can squeeze through a gap the width of a fountain pen.

Damage Caused

House Mice breed rapidly and a colony can soon build up. They are particularly attracted to chocolate, nuts, seeds and cereal substances, but contrary to belief are *not* particularly keen on cheese.

54

House Mouse nest in a piano.

Mouse body smears on the corner of a wall.

Food packs fouled by mice.

Mice nibble into profits.

Mouse damage to electric wiring.

In common with all rodents mice have a compulsive need to gnaw in order to keep their incisor teeth worn down to a constant length. Electric cables, water pipes and gas pipes, packaging and woodwork may all be seriously damaged by mice and many instances of electrical fires and floods have been recorded as a result of mouse activity. They contaminate far more food than they consume and are capable of carrying many diseases particularly *Salmonella* food poisoning.

Body smears show where a rat moved across this roof.

A mouse will shed about 70 droppings every 24 hours and has no effective bladder and so its 'dribble is worse than its nibble'.

The small dark droppings and evidence of gnawing are the most obvious signs of infestation but where a colony is well established, dark smears of grease from their body hairs may be apparent on walls and pipe runs.

Control Measures

Mice are erratic, sporadic feeders, nibbling at many sources of food rather than taking repeated meals from any one item. They do not need free water to drink as they normally obtain sufficient moisture from their food.

Because of these habits, traditional baiting techniques and trapping are often unsatisfactory and a combination of rodenticides may be necessary in addition to mouseproofing as far as possible, all means of entry. Holes should be blocked with wire wool embedded in cement. Pipes and cable ducts should fit closely in any partition they pass through, without leaving gaps.

'Break-back' traps can be useful for dealing with the odd one or two mice but must be correctly placed with the treadle at right angles to the wall beside which mice tend to run. Experimental work with a known population of mice has shown that only about 50 percent of them could be caught by traps alone. Bait the traps with chocolate, nuts or dried fruit or use the new type that needs no bait.

The domestic cat is seldom effective at mouse control inside buildings as the mice are usually in areas inaccessible to a cat. Cats are not acceptable in areas where food is prepared commercially.

Since World War II, the most widely used and effective modern rodenticide has been warfarin. This is an anticoagulant drug which prevents blood clotting and mice feed on it without suspicion, growing weaker until they die, apparently without pain. Only a very small proportion of the chemical is mixed with the bait, usually oatmeal, so there is little danger of humans or pets consuming a lethal dose accidentally.

In most urban areas of Britain and in some other countries, however, mice have now become immune to warfarin and different types of mouse

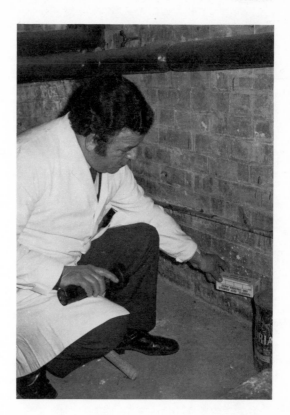

Placing a box containing
mouse bait.

killer based on alphachloralose, difenacoum and calciferol have been
developed. These are quick, humane and safe to use indoors as directed
on the instructions. Alphachloralose sends the mice to sleep and induces
hypothermia so they die without regaining consciousness. Calciferol
upsets the calcium balance in the body inducing hyper calcaemia. Other
rodenticides such as lindane contact dust which adheres to the fur and
feet of the mouse can be used by pest control contractors or local
authority public health departments. Proprietary rodenticides are sold
ready-mixed in a suitable bait, coloured for identification and should
not be added to other bait material.

Commercial premises should have a regular pest control contract to
prevent mice problems.

Related Species

The Long-tailed Field Mouse or Wood Mouse *Apodemus sylvaticus* often comes indoors in Autumn and Winter. It is brown above, white beneath, with larger ears and more prominent eyes than the House Mouse.

A destructive pest of seeds and fruit in the garden, it also invades beehives to eat honeycomb, and indoors will eat seeds, bulbs and cereal foods.

In parts of Northern Europe, the Bank Vole *Clethrionomys glareolus* sometimes enters houses. It is distinguished by its blunt muzzle and short tail and reddish-brown colouring. All are susceptible to the rodenticides for house mice and to warfarin or difenacoum based products.

Parasites

'So, naturalists observe, a flea
Hath smaller fleas that on him prey;
And these have smaller fleas to bite 'em,
And so proceed ad infinitum'.   Jonathan Swift

Bed Bugs

Improved environmental hygiene has reduced the Bed Bug to something of a rarity in many countries but it still exists in poor housing, old flats and in some boarding houses, holiday camps and hostels. In the UK it is one of the insects which qualifies for causing 'verminous conditions' according to the Public Health Act and a local authority is empowered to disinfest such premises or give notice to the owner to have them disinfested at his own expense.

Identification

The Bed Bug *Cimex lectularius* has a brown round flat body about 4 to 5mm (about ⅛in) long that enables it to hide in the narrowest of crevices. Its wings have degenerated and virtually disappeared. Both the nymph and adult have piercing mouthparts and feed on blood. Their bodies become distended with blood when they are feeding.

Life Cycle

The female lays one to five pearly eggs a day for up to two months in suitable cracks and five nymphal stages are passed through until the insect becomes adult. The eggs are stuck to rough surfaces of bedsteads, plaster beneath wallpaper or in the cracks inhabited by the adults. In warm weather they hatch in about 7 to 10 days. The adult may live from 9

Bed Bug.

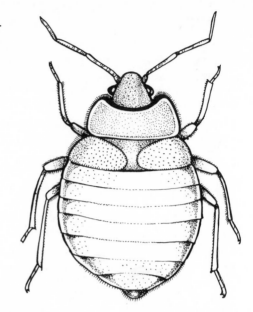

to 18 months at room temperature and can survive considerable periods of starvation. The adults wander about seeking suitable harbourages and may spread to other rooms or adjacent flats. In unheated rooms they are inactive in winter but in warm countries or heated rooms they reproduce and feed all the year round.

Damage Caused

Overcrowded, dirty rooms may support large numbers of these flat brown insects behind peeling wallpaper, in cracked plaster or woodwork or in bed frames, headboards and mattresses. In blocks of flats they move between apartments along the service ducts or heating ducts. The bugs have piercing and sucking mouthparts for their role as human parasites and emerge at night in an occupied room, causing irritating bites to the human host. An unpleasant smell is produced by them and they are often introduced or conveyed from one house to another in infested bedroom furniture, cases, or clothing that has been hung against an infested wall.

Some people are very sensitive to the bite of the Bed Bug, others are not. It can produce severe itching and a small, hard whitish swelling which distinguishes it from the typical red spot of a flea bite.

Control Measures

Thorough cleaning of floor, walls, beds and bedding is essential, with special attention to scraping out cracks in floorboards and under skirting. Fumigation of infested premises has largely been replaced by the use of persistent insecticides. If a lindane spray is used, a liberal application of about $4\frac{1}{2}$ litres (half a gallon) of the diluted preparation to a bedroom is called for, paying particular attention to mattresses, bedsteads, loose paper, the backs of furniture and pictures and skirting boards. Alternative insecticides, such as malathion or iodofenphos as a wettable powder, are used successfully by contractors. Fumigation of bedding and upholstered furniture, however, is often necessary and this can be carried out by a fumigation contractor or a furniture preservation service. Persistent infestation should be dealt with by the local Public Health Authority or a pest control company.

Related Species

Bats and birds are hosts to several *Cimex* species which in certain circumstances may migrate from them into buildings. Starlings' and Pigeons' nests are often infested with bugs and these will occasionally transfer their attentions to people. The Martin Bug similarly may invade buildings from nests of the House Martin.

Lice

The insects generally referred to as lice are animal parasites belonging to one of two groups—biting lice or sucking lice. The Book louse is not a parasite nor is it related to the parasites (see page 98). For our purpose they are best divided in Dog Lice and Human Lice.

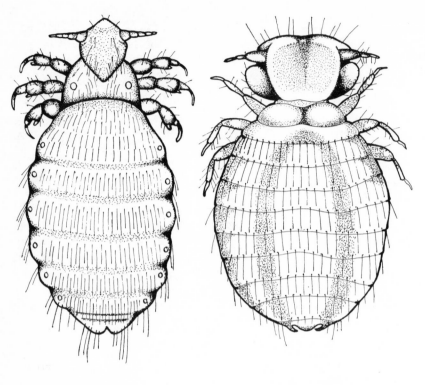

Dog-sucking Louse.   Dog-biting Louse.
(After Hickin)   (After Hickin)

Dog Lice—Identification

The Dog-biting Louse *Trichodectes canis* is the commonest species found indoors where dogs are present and is even more prevalent than the Dog-sucking Louse. In common with other lice, the small brownish body of the adult, which is about 1.5 mm ($\frac{1}{16}$ in) long, is flattened from top to bottom. This distinguishes it from the flea which is flattened from side to side.

Life Cycle

The eggs ('nits') are cemented to the base of a hair on the dog and the nymphs and adults feed on dead skin, usually on those areas of the body behind the ears, on the flanks and at the base of the tail. There is no larval stage, the nymphs that emerge from the eggs simply grow until they reach maturity and the life cycle is completed on the dog's body.

The Dog-sucking Louse *Linognathus setosus* which may infest the same areas of long-haired dogs, is of importance not only because of the irritation it causes, but because it is the means of passing on the dog tapeworm.

Control Measures

Veterinary insecticidal powders should be used in accordance with the manufacturer's instructions to disinfest affected animals.

Bird lice may also be accidentally introduced to a house from nests on the building, perhaps on a chicken's feathers or by the family pet but they soon die in the absence of their natural host.

Human Lice—Identification

Two forms of the Human Louse *Pediculus humanus* are recognised, depending on whether they infest the head or the body.

A separate species, the Crab Louse *Phthirus pubis* confines its attentions to the pubic hair, hence the nickname 'Papillon d'amour'.

Both species are sucking lice, with mouthparts adapted for piercing skin and sucking blood. Adults are greyish, up to 3 mm ($\frac{1}{8}$ in) long.

They are wingless greyish insects with strong clawed legs. In *Pediculus* the legs are all about the same and the abdomen is twice as long as it is broad, but in *Phthirus* the second and third pairs of legs resemble a crab's claws and the abdomen is broader than it is long.

Life Cycle

The pearly oval eggs, known as 'nits' are cemented to hairs or to clothing

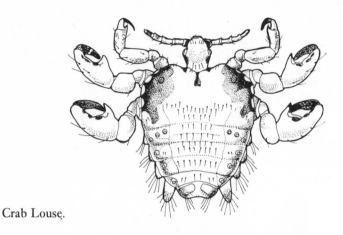

Crab Louse.

fibres. There are three nymphal moults and both nymphs and adults feed on human blood. The life cycle is completed in about eighteen days, nine days for the egg to hatch and nine days for the nymphs to mature. The Body Louse is more robust than the Head Louse.

Damage Caused

In the past the Body Louse has been responsible for severe epidemics of Typhus and Trench Fever.

The bites of lice cause irritation and people scratching themselves to relieve this may introduce secondary infections such as impetigo. Considerable feelings of shame and anxiety are also induced by having lice although they do not necessarily indicate bad hygiene.

Improvements in environmental and personal hygiene have greatly reduced infestation by Body Lice and Crab Lice, but the incidence of Head Lice has recently been rising again in schools, especially among young girls or where long hair is permitted, to the extent that up to a million schoolchildren in Britain are said to be infested.

The British Education Act (1944) empowered the Medical Officer of Health or his agents to inspect children in local authority schools and to have children bearing lice disinfested within 24 hours. Similar provisions for dealing with infested persons are made by the Public Health Act.

Control Measures

If lice infestation is suspected, advice should be sought from a doctor. The fact that lice are contracted does not imply that one is dirty, but the sooner the trouble is reported the sooner the source of infestation can be traced and dealt with. Head lice have become resistant to some previously effective insecticides but modern clinical preparations applied to the affected area and clothing can soon clear up the problem.

Crab Lice may be treated by shaving the pubic hair and applying carbaryl or lindane powder liberally to any affected clothing. Carbaryl and malathion lotions or shampoos control head lice.

Fleas

Fleas are flightless parasites of warm-blooded animals, characterised by a side to side flattening of their bodies and a well developed ability to jump. The lateral compression of the body helps the flea to move quickly through fur or clothing. Each species prefers to suck blood from a particular species of animal but when the preferred host is absent each will bite others, including man, causing considerable local irritation to susceptible persons. At one time fleas were referred to as 'Jumping dandruff' and were used to entertain in Flea Circuses.

The three species which may occur in dwellings are the Human Flea *Pulex irritans*, the Cat Flea *Ctenocephalides felis* and the Bird Flea *Ceratophyllus gallinae*. The Cat Flea readily bites man and is the species most often found in infested buildings. The relatively uncommon Dog Flea can carry eggs of the dog tapeworm in its gut, which may develop if the dog swallows the flea in an attempt to allay irritation. According to a Rentokil survey covering five years, requests to deal with fleas outnumbered those for Bed Bugs by 3 to 1. Approximately 80 percent of infestations dealt with referred to Cat Fleas. The Human Flea also occurs on foxes and badgers.

Identification

Fleas are wingless, with laterally flattened bodies 2–3mm long and long

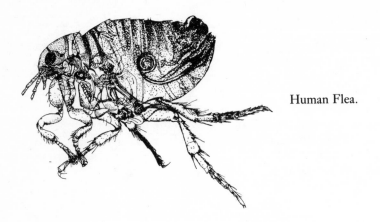

Human Flea.

muscular legs. They are reddish-brown in colour with a helmet-shaped head which in many species bears a comb of dark spines that help in the identification of the species.

Life Cycle

Eggs are laid in the fur, feathers, clothing or sleeping places of the host. They are very small, pearly-white and oval in shape, just visible to the naked eye and several hundred may be laid by one adult in batches of four to eight. The egg hatches in about a week and in summer in Britain the complete life cycle takes about a month. At winter temperatures it is much longer. There is a peak of flea infestations in late Summer, falling away in Autumn. This is the time when most biting adults are present. The larvae are white, legless, thread-like creatures with a number of backward-directed bristles. The grubs' development takes place away from host animal, in or under its nest or sleeping place. The larvae feed on the faeces of adult fleas or other organic debris and may persist in the fluff around or under carpets especially in heated buildings.

The pupae in silken cocoons are at first white but turn brownish or grey before the adults are ready to emerge. The adults do not break out, however, until vibrations indicate the presence of the host near the pupal site—usually the host's sleeping quarters.

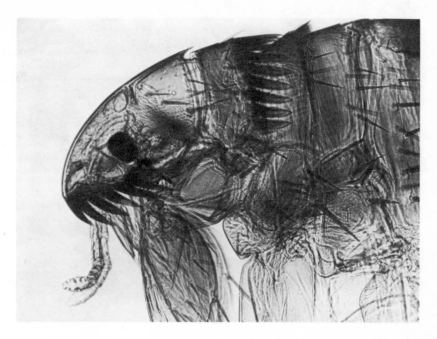

Head of Cat Flea.

Damage Caused

Because adult fleas do not emerge until they detect a host, a person visiting an old unoccupied house may receive the attention of a large number of fleas in a very short time. If this happens it is wisest to strip off clothes in an empty bath because the fleas cannot climb its smooth sides.

The obvious damage caused by the flea is its irritating bite which leaves a small red spot. Both males and females suck blood and both the Cat Flea and the Bird Flea will bite humans.

The geographical distribution of fleas as reported to UK local authorities shows them to be a problem in high density cities and in many holiday resorts.

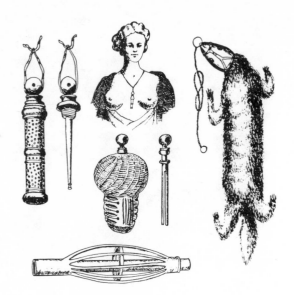

18th-century flea traps
worn beneath clothes to
attract any fleas on the
person.

Control Measures

Treatment is aimed at controlling the infestation by application of insecticide to the breeding sites and the bedding where possible. Control should be carried out in two stages:

Thoroughly clean the floor and bed, preferably with a vacuum cleaner. Scrape out dirt from between cracks in the floorboards and from under skirtings. If the living quarters are dirty enough to support flea larvae, then these areas too must be cleaned. Have all bedding cleaned by washing or dry-cleaning, and if possible, upholstered furniture and mattresses should be fumigated.

Insecticidal treatment of the bed frame, the floor area under and around the bed, and any other infested areas, should be done by spraying and dusting with an insecticidal powder or aerosol.

Always treat infested cats and dogs with a special veterinary flea powder in accordance with the instructions and use the same material to treat the clean bedding. Burn old infested bedding, treat crevices in flooring or walls and remove old birds' nests from the eaves or loft.

For rapid clearance of an infested building a pest control company may use an insecticidal fog or spray with iodofenphos insecticide. Fumigation is rarely necessary.

Related Species

It was the Rat Flea *Xenopsylla cheopis* that spread the Black Death through Europe in the thirteenth century and which caused The Great Plague of London in the seventeenth century. Although Bubonic Plague is still endemic in some parts of the world, rodent fleas in Britain are today rarely found in buildings.

Most wild animals, from foxes and rabbits to hedgehogs and bats have their own fleas, but these are rarely carried indoors.

Worms

Parasitic worms are a veterinary or medical problem, but some forms are transmitted by rats, mice and insects.

Identification

Threadworms are fairly common in dogs, cats and children. The thin white wrigglers emerge from the bowel at night to cause intense itching in young children. Roundworms are like small grey earthworms. Tapeworms have long, flat segmented whitish bodies with a narrow hooked head with which the worm fastens itself inside the gut.

Life Cycle

Eggs or egg-bearing segments pass out of the bowel or are laid on the buttocks by emerging worms and can be transferred by dirty hands to the mouth and swallowed.

Damage Caused

Irritation, loss of energy, in animals a depraved voracious appetite and general debility.

Control Measures

Consult a veterinary surgeon or doctor who will prescribe effective pills or worming tablets. Dichlorophen is effective against dog or cat tapeworm.

Insist that children wash hands immediately after using the lavatory and before meals. Wash all fruit and vegetables thoroughly before eating them. Keep fingernails short.

Flying Insects

'God in his wisdom
Made the fly
And then forgot
To tell us why'   Ogden Nash

Flies

Under this heading are included species that are generally referred to as flies by the layman. Other species of the order Diptera, also characterised by having a single pair of wings, are described under their common names such as Gnats or Mosquitoes.

Flies are probably the most commonly tolerated pests, yet they are among the most loathsome. The most familiar and typical are the two species of House Fly.

Adult House Fly.

73

Identification

The Common House Fly *Musca domestica*, is almost too familiar to need description. The adult is 7–8 mm ($\frac{5}{16}$ in) long, grey in colour with black stripes on the thorax, the body is covered in short hairs and with two veined membranous wings. The large compound eyes take up most of the head. The Lesser House Fly *Fannia canicularis* is rather smaller, the thorax having less distinct stripes. It is most easily distinguished by its flight, which is relatively slow but with sudden changes of direction as it cruises around light fittings or other hanging objects. The Lesser House Fly is now the commonest of the two species in buildings.

Life Cycle

An adult House Fly may lay 600–900 eggs in batches of 120. Both species lay eggs in decaying organic matter such as rotten vegetables or household refuse and these may hatch within 24 hours. The white, legless larvae or 'maggots' burrow into the food and, in a warm summer, may complete their development in a week. When fully fed the larvae crawl away from the food to pupate in nearby soil or refuse, emerging as adults in one to four weeks. The answer to the question 'Where do flies go in the Winter time?' is that a few adults hibernate indoors, and others pass the Winter in the pupal stage. In the high temperature and humidity of some restaurant kitchens flies may be active throughout the year.

The adult House Fly lives from four to twelve weeks. Breeding stops in Autumn. It is commonest and most active in the hottest months, and forages among filth in dustbins and refuse tips where it may pick up a wide variety of bacteria in its gut, on its feet and attached to its body hairs.

Damage Caused

Over three million bacteria may be present on the body and legs of one fly. The fly may then land on sugar, icing, meat, or other human food. Because it only has a sucking mouthpart for feeding, the fly vomits saliva on the food and treads this pollution into the surface until it is sufficiently liquid to be sucked up. Each foot is equipped with a special cushion (the

A fly's foot, magnified, showing hairs and the sticky pulvillus at the end.

pulvillus), covered with sticky glandular hairs which enable the fly to walk upside down, but which also gather bacteria from refuse or faecal matter among which it forages and breeds. The so-called 'fly spots' are regurgitated material from the gut of the fly.

As a consequence of these habits, flies transmit a wide variety of gastro-enteric illnesses such as typhoid and summer diarrhoea. In warm countries flies are responsible for spreading diseases such as dysentery, ophthalmia, diphtheria, and cholera.

The eggs of parasitic worms can also be carried on a fly's feet. It has been said, with justification, that 'If you followed a fly for a day, you wouldn't eat for a week'.

An electric ultra-violet
light trap to electrocute
flies.

Control Measures

Control of all houseflies is exercised in three directions. Scrupulous hygiene and prompt disposal of all food residues, empty tins and bottles is necessary to deprive them of possible breeding sites. Every disposal bin or refuse bag should have a close fitting lid and be thoroughly emptied at regular intervals. It is a good idea to spray or dust the inside of the dustbin with an insecticide during the summer. Good housekeeping will generally minimise the risk of large scale breeding.

Fly screening of windows, especially kitchen windows and fanlights, is fairly easy with flexible plastic flyscreens which use Velcro to fasten them to the window frames. Adult insects that get indoors are best dealt with by a quick knockdown insecticide such as an aerosol containing pyrethrins. For more permanent protection, you can ask a pest control company to apply a long-lasting organo-phosphorus insecticide such as diazinon or one of the new pyrethroids to pendant fittings and surfaces on which flies most often alight. For canteens, food factories and other industrial areas, electric ultra-violet fly killing devices are available which attract and electrocute flies, dropping their bodies into a catching tray.

Automatic electrically operated dispensers of insecticide may be used in unoccupied rooms, but those constantly releasing lindane are not recommended for any living accommodation that is constantly occupied.

Impregnated strips of plastic giving off vapours of the organo-phosphorus insecticide dichlorvos are also sold for domestic fly control, but it is preferable to use controllable devices from which the rate of emission can be adjusted and which can be closed down when not required.

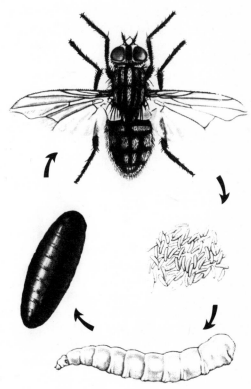

Life cycle of House Fly.

If large numbers of flies such as Cluster Flies accumulate in a loft or roof space, it is a good idea to use a lindane smoke generator which leaves an even deposit of insecticidal particles on all horizontal surfaces. A vacuum cleaner may be used if flies are accessible. Suck some insecticide powder into the bag first, then vacuum up the flies and leave them in the bag for a while. Sticky flypapers can be hung up to catch flies for those who prefer not to use chemical methods.

Related Species

Other flies of domestic importance include the Blowflies, usually described as Bluebottles or Greenbottles *Calliphora* sp., the Cluster Fly *Pollenia rudis*, the Fruit Fly *Drosophila* sp. and the Crane Fly or 'Daddylonglegs' *Tipula* sp. Filter Flies and Seaweed Flies *Coelopa frigida* can be a local problem at the coast or near sewage works.

The Blowflies are so called because of their habit of 'blowing' or depositing their eggs on exposed meat. They include the Bluebottle *Calliphora erythrocephela*, characterised by a shiny blue body, the two Greenbottles *Lucilia sericata* and *Lucilia caesar*, and the Flesh Fly *Sarcophaga carnaria*, which is black and grey. All are much larger than the House Fly and have a loud buzzing flight.

These species like bright sunlight, but in Summer they often enter houses in search of breeding sites. They are frequently encountered around the dustbin on a hot day. The eggs are laid on animal matter, meat or carrion, except in the case of the Flesh Fly which deposits small maggots.

Up to 600 eggs may be laid by one Bluebottle and will hatch within 48 hours at 68°F (20°C). At such a temperature the maggots develop fully in about ten days. They burrow into meat as they feed on it. Like the House Fly, the larva buries itself several inches in soil before pupating and emerges from its brown pupa in about ten days.

Cluster Flies and Swarming Flies often invade attics, roof voids and the upper rooms of houses in Autumn, seeking sites for communal hibernation. The cluster of flies becomes active only when there is a rise in temperature. It is almost impossible to prevent an invasion by these flies if they choose a particular house and they may return each year. One species, *Pollenia rudis*, is parasitic upon a species of earthworm. The tiny

Cluster Fly.

red-eyed Fruit Fly *Drosophila melanogaster*, is one of several species associated with fermenting liquids or rotting fruit. They are also known as wine flies, yeast flies or vinegar flies. They have a characteristic slow, hovering flight with their abdomen hanging downwards.

The Crane Fly or Daddylonglegs *Tipula paludosa* is the adult form of the 'leatherjacket' which is a pest of lawns. It does no harm but is often confused with the mosquitoes because of its large size, long slender body and narrow wings. It is larger than a mosquito and lacks that insect's long proboscis.

A 'House Fly' that bites is probably a Stable Fly, especially if cattle or horses graze nearby or the house is near a seaweed-covered beach.

The Stable Fly *Stomoxys calcitrans* breeds among straw or decaying vegetation such as grass cuttings or seaweed, feeds on the blood of cattle or horses and will bite people.

79

The adult has a slender proboscis projecting forward from the head, several dark spots on the abdomen, and at rest holds its wings at an angle to the body instead of projecting them almost straight backwards as the housefly does.

Breeding areas may have to be dealt with by a contractor or local authority using insecticidal fogging equipment.

If garden compost is infested spread it thinly as the larvae are susceptible to drying out.

Cheese Skipper

Meat and cheese are still sometimes found infested by slender grubs that shoot themselves along by a 'skipping' motion. At one time these creatures were said to form the tastiest parts of a cheese!

Identification

The adult Cheese or Ham Skipper *Piophila casei* is a small shiny black fly, with reddish eyes.

Life Cycle

Eggs are deposited on the surface or in crevices in cheese or meat or their wrappings. The larvae burrow into the food and feed for between 5 and 15 days. It is only the third instar larva that 'skips' by curving its body into a ring and suddenly releasing itself.

When ready to pupate the larvae emerge from the food to seek a dark dry crevice and form a reddish puparium.

Damage Caused

Lean meat, ham, bacon and mature cheese can be ruined by this insect and the larvae can cause internal irritation if accidentally eaten.

Window Gnat.

Control Measures

Good housekeeping will generally eliminate the Cheese Skipper. Leave no traces of grease or cheese crumbs on larder shelves and keep meat and cheese products covered or in a cool place. An aerosol of diazinon or pyrethroid sprayed into all possible pupation sites after removal of food and thorough cleaning will protect against immediate re-infestation.

Gnats

A number of species of non-biting gnats sometimes enter houses, especially in the neighbourhood of sewage works.

The insect normally referred to as a gnat is in fact the small mosquito *Culex pipiens*, which does not bite humans. However, it looks very much like *Culex molestus* which does bite, and painfully. It breeds in stagnant water such as water butts and a film of paraffin sprayed over the surface of such water will kill the larvae.

The Window Gnat *Anisopus fenestralis* is also sometimes confused with the mosquito but has more rounded wings and abdomen. The larvae have caused damage to home made wines. Eggs are laid on rotting fruit or vegetables or other moist food.

Gnats usually remain still for long periods around windows and are easily knocked down by pyrethrin aerosols.

Midges

Tiny dark grey or black flies of several species most of which do not bite but are almost indistinguishable from those that do! Americans call biting midges 'no see-ums'.

Identification

Small, about 2 mm ($\frac{1}{12}$ in) long with hair-fringed wings, 'humped' thorax and 'dancing' flight or 'hopping' movements. Variously known as Filter Flies, Owl Midges or Bathroom Flies, the commonest is *Psychoda alternata*.

Often abundant in Spring and Summer around sewage works.

Culicoides obsoletus, a biting species, folds its wings flat along its back whereas the non-biters rest with wings held tent-wise over their bodies.

Life Cycle

Eggs are laid in slime around drains and outlet pipes and the larvae live in this, feeding on algae, bacteria and organic waste until they mature, pupate and adults emerge.

Damage Caused

Beneficial in sewage purification filter beds but a nuisance indoors. Bites of *Culicoides* are intensely irritating.

Control Measures

Remove slime from possible breeding areas. Fly screens over open windows and sticky flypapers or one squirt of a pyrethrin aerosol in rooms will kill adults. Apply insect repellent to exposed skin if venturing into known midge-infested areas.

Mosquitoes

The long narrow body of the mosquito, with long pointed wings held along it when at rest, is familiar to most people.

The eggs are always laid in water, often in water butts, ponds or guttering and the small, wriggling larvae hang from the surface.

There are about 30 British species but only three or four commonly enter houses. All are smaller than the Crane Fly with which they are sometimes confused.

Identification

There are two main groups of mosquitoes, the *Culicine* and the *Anopheline*. The adults have prominent eyes and proboscis on the head, a thick thorax and long thin abdomen. They have long thin legs and strongly veined wings with scales at the edges. The *Culicine* mosquitoes hold their abdomen parallel with the surface on which they perch and have no wing spots.

The *Anopheline* mosquitoes hold their abdomen at an angle to the surface and most species have spots on their wings.

In Britain, *Anopheles maculipennis* is fairly common indoors, especially in marshy areas around the coast or in estuaries. It is brownish in colour with small spots on the wings. It bites readily, especially at dusk. It is the commonest European carrier of malaria.

The largest British species is *Theobaldia annulata* which has dark spots on the wings and white bands on its legs (see page 84).

83

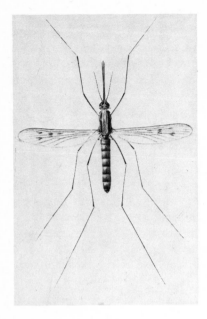

Mosquito.

Life Cycle

Mosquitoes start life as eggs laid on the surface of water. The larvae or 'wrigglers' have a prominent head, a thick thorax and a tuft of bristles on the abdomen. They obtain air by hanging from the surface of the water, the Anopheles species lying parallel beneath the surface, other types hanging head downwards.

The eggs are laid in batches of about 50–200, hatch in a few days and the larval stage takes from 4 to 10 days in most species while they feed on minute animal and plant life filtered from the water. The pupal stage also stays in water and is usually described as 'comma shaped' and the adult emerges in a day or two.

Adult mosquitoes hibernate in dark, sheltered places such as cellars, cow sheds, outbuildings or in hollow trees. Only the females are literally blood-thirsty, requiring a meal of blood before they can lay fertile eggs.

Damage Caused

The most apparent form of mosquito damage is their bite. Most mosquitoes feed on blood and many of them bite man. Some people have bad reactions to mosquito bites with severe itching and swelling of the bitten area.

In many parts of the world malaria is transmitted by various species of Anopheles mosquito. The diseases of yellow fever and dengue fever are carried by *Aedes aegypti* and other species are vectors of the organisms causing elephantiasis and encephalitis. In Ethiopia in 1965 there were 300,000 cases of yellow fever with 30,000 deaths; worldwide cases of malaria are in hundreds of millions every year.

Control Measures

World eradication programmes involve vast sums spent on insecticides.

Adult mosquitoes can be kept out of homes by fitting flexible flyscreens to windows. In tropical regions, a mosquito net should be used for sleeping under at night.

Breeding sites such as water butts or stagnant pools should be treated by placing a few drops of paraffin or kerosene on the surface to cut off the air from the larvae and pupae.

Bird baths should be emptied regularly to prevent mosquitoes breeding. Adults can readily be killed by a pyrethrin or other insecticidal aerosol.

In the case of persistent serious infestation by mosquitoes, inform your local public health department.

Related Species

See under 'Gnats' (page 81) for the common species *Culex pipiens*.

Wasp nest in a roof.

Wasps

With their yellow and black striped bodies wasps are a major nuisance in late Summer when foraging workers enter kitchens and any place offering access to sweet food. The late Summer months are the peak times of wasp activity.

Identification

The two species most often found indoors, the Common Wasp *Paravespula vulgaris*, and the German Wasp *Paravespula germanica*, look identical to the casual observer and have a similar life cycle and habits.

The workers are 10–15 mm ($\frac{1}{2}$ in) long, the queen is about 20 mm ($\frac{3}{4}$ in) long, and all have the bands of yellow and black across their bodies and the distinctive narrow 'wasp waist'.

86

Life Cycle

Towards the end of the summer a queen wasp lays a number of unfertilised eggs which provide male wasps to mate with new queens. Only new queens survive the Winter in hibernation and next Spring each selects a nest site in a bank or in the ground, or in the rafters of a shed or house. The nest is made from wood scraped from a post or fence and worked with saliva into a paper-like consistency. Inside are a number of combs of hexagonal shaped cells in which the eggs are laid. The grubs are fed on dead insects by the queen and finally develop into worker wasps in three or four weeks. Workers may forage up to three miles from the nest in search of food and one nest may produce 30,000 wasps in the course of a season.

Hornet.

A Rentokil 'Rentoflash' with its 'kill' of wasps.

Damage Caused

Wasps become a nuisance because they have a sweet tooth at one end and a painful sting at the other. They generally cause more anxiety than harm but the sting is painful and several deaths occur each year from anaphylactic shock among people who are exceptionally susceptible to its effect. The constituents of the venom are a histamine and various toxins and the best antidote is afforded by an anti-histamine cream after making sure that none of the sting mechanism is left in the skin. The preparation sold in small aerosols as 'Wasp-Eze' contains a local anaesthetic as well as an antihistamine and is most effective at relieving the pain of any insect sting or bite if applied promptly.

Control Measures

Household aerosols or insecticidal sprays will knock-down adult wasps

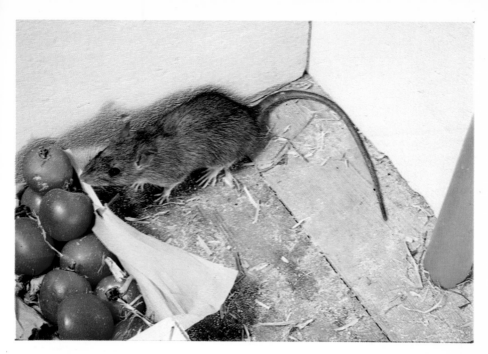

Black Rat
Brown Rat

Bluebottle
House Mouse

Housefly on bacon
Housefly pupae

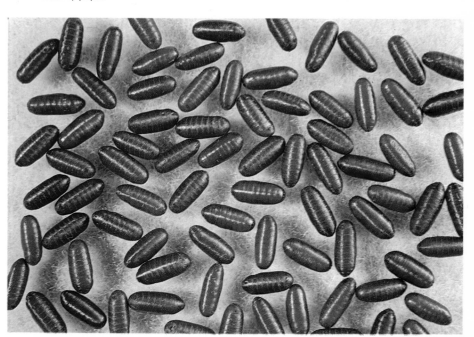

Nymphs of German Cockroach
Blow Fly 'maggots'

Larder Beetle
Bed Bug

Brown House Moth
Carpet Beetles: adults and larvae

Common Furniture Beetle
Furniture Beetle pupa in pupal chamber

Worker Termites
Dry Rot fruiting body

and a nest in a bank or wall can be dealt with by puffing a proprietary Wasp Nest Killer based on carbaryl into the entrance. In an inaccessible roof space lindane smoke generators or chlordane spray may be effective. Treat nests at dusk when the wasps are safely inside, wear gloves as a precaution and leave the vicinity of the nest quickly. Some public health authorities will remove wasps' nests, or a pest control company will do the job for a few pounds.

Wasps like other flying insects can be kept out of kitchens, caravans or food areas by judicious use of flexible fly screens fitted to window or fanlights.

Industrial catering or food production areas can be protected by electrical devices which attract wasps or flies to an ultra-violet light, electrocute them and collect the bodies in a catching tray. For some premises baits of slow acting insecticide can be placed around the perimeter to attract worker wasps which then carry the insecticide back to destroy the nest.

Related Species

The Hornet *Vespa crabro* is relatively rare but occasionally enters buildings. It is much larger than either wasp or bee and has brown rather than black head and thorax.

Crawling Insects

'Go to the ant thou sluggard'
Proverbs

Ants

Identification

Ants in the pantry are no joke. They are a source of nuisance and anxiety. The Black Garden Ant *Lasius niger* is the species most often encountered indoors. All ants have a characteristic narrow 'waist' between the thorax and the abdomen and their antennae are jointed or 'elbowed'. The adult worker Black Garden Ant is 3 to 5 mm (about $\frac{1}{8}$ in) long. Winged males and females emerge in July and August.

Around buildings small piles of fine earth or mortar brought up by the ants from beneath paving stones indicate the openings of the underground nests.

Life Cycle

The so-called 'ants eggs' sold for feeding to pet fish are not the eggs but are the ant pupae. The real eggs are very small, hatch in 3 to 4 weeks and the first larvae are fed by the queen with a special secretion until they pupate and emerge as worker ants. These and subsequent workers forage for food outside the nest, and feed and tend the larvae hatching from new batches of eggs laid by the queen.

The food they collect includes dead insects, sweet materials and the 'honeydew' secretion which they get by herding and 'milking' aphids on plants.

The workers are sterile females but winged sexual forms are produced which 'swarm' for the mating flight during which new queens are fertilised. The males die but the queens bite off their wings and spend the Winter underground, ready to start laying eggs for a new colony in the Spring.

Damage Caused

Although the Black Garden Ant is not known to transmit disease, any food rich in sugar is intensely attractive to the workers and they will cluster all over it, entering houses through the smallest cracks and crevices to reach it.

Control Measures

Ants can be easily controlled if all their nests are found but these are often inaccessible and the foraging worker ants have to be dealt with wherever they are encountered.

Black Garden Ants which invade the house should be traced to their nest outside, which is usually easily located, and boiling water can be poured into the entrance hole as a quick 'first aid' measure. A spray or dust containing lindane, carbaryl or boric acid is more effective and more permanent, but these substances should not be used near food. Proprietary baits containing boric acid can be effective and attract the foraging workers which destroy the whole colony by feeding it to the queens and young in the nest. Effective protection is also given by painting with insecticidal lacquer or spraying with an aerosol of diazinon the points of entry and pipe runs, skirting, etc. along which the insects run. Boric acid powder pumped into wall and floor cavities gives long term protection. If the problem persists, contact your Environmental Health Department or a pest control company.

Related Species

Pharaoh's Ant *Monomorium pharaonis*, an imported tropical species, cannot live outdoors in Britain but has established itself in many large,

Left: Female German Cockroach with egg capsule. *Right:* Female American Cockroach with egg capsule.
These photographs are for identification: they are not a true indication of size. The American Cockroach is in fact three times the size of the German Cockroach.

centrally heated buildings where there is high humidity and plentiful food. It is a major pest in the Netherlands and appears to be increasing throughout Europe and the U.S.A. It is a tiny yellowish brown insect only 2 mm ($\frac{1}{12}$ in) long which can penetrate into the smallest spaces and is often difficult to eradicate without specialist help, especially in large hospitals or hotels. One way of tracing this pest to its hiding place is to place baits of raw liver and then follow the ants' route after feeding. Iodofenphos or boric acid baits have been used successfully by pest control contractors to control these ants, using raw liver as the attractant, as Pharaohs ants are protein feeders. The Argentine Ant *Iridomyrmex humilis* is occasionally found in centrally heated buildings including hospitals in Britain and more frequently in warmer countries. Similar in appearance to Pharaohs Ant, it prefers sweet food. Both species can be carriers of cross infection in hospitals and in one survey nineteen pathogenic organisms were found on Pharaohs ants taken from nine hospitals.

Cockroaches

Although more common in industrial and catering premises than in homes, the cockroach may be found in older houses and flats, usually becoming apparent in bathrooms and kitchens where there are warm moist, favourable clustering sites around pipes, stoves and sinks. It is also common in multi-storey buildings such as hotels and apartment blocks where central heating ducts provide warmth, humidity and access to many rooms.

The two species most often encountered in Britain, the Common or Oriental Cockroach *Blatta orientalis* and the smaller German Cockroach (or 'Steamfly') *Blattella germanica* differ somewhat in their habits, and control methods may have to be adapted accordingly. The large chestnut brown American Cockroach *Periplaneta Americana* also occurs, often transported from ships.

They are all nocturnal in habit, emerging at night from the narrowest cracks to forage for any of a wide variety of human food. They contaminate more than they consume, polluting everything with a foul 'roachy' odour which is persistent. In the Tropics the number of pest species indoors increases but their biology and habits are similar.

Identification

Cockroaches are large insects, the German adults being up to 14 mm (about $\frac{1}{2}$ in) long, the Oriental species reaching 24 mm (nearly 1 in).

Adults are shiny dark brown almost black in the Common or Oriental Cockroach, and light brown with two dark brown stripes on the thorax in the German species. All species have characteristically long whip-like antennae, flat bodies and rapid, jerky movements.

Adult Common or Oriental
Cockroach.

Egg capsule of Common
Cockroach.

Loaf infested by American cockroaches.

Life Cycle

Eggs are laid in dark brown capsules about 12 mm ($\frac{1}{2}$ in) long. Each capsule of the Common Cockroach contains sixteen eggs and is deposited in a warm sheltered place near a food supply. Infestations of cockroaches often start from accidental introduction of egg capsules into buildings, in food boxes, beverage crates, packaging materials and other containers. The German Cockroach carries its egg capsule with it until the twenty or more young are ready to hatch. There is no larval stage in cockroaches, the young emerging as 'mini-cockroach' nymphs which resemble the adults and which grow to maturity by a series of moults. Length of the life cycle depends on the extent of the warmth and humidity. These species do not normally live out of doors in Britain.

Damage Caused

The vomit marks and excreta of cockroaches as they forage over food or

Using a power-operated sprayer to apply insecticide against cockroaches in heating ducts under a hospital.

food storage areas at night after emerging from their daytime hiding places, spoil food and may taint it with an objectionable and persistent smell. The cockroach has been shown to be capable of carrying a wide range of disease-causing organisms and must be considered a possible vector of food poisoning in catering premises and cross-infection in hospitals.

Control Measures

The basis of effective cockroach control, is to place insecticide where the insect hides and it often needs specialist equipment to deal with cockroaches in their inaccessible harbourages. Formulations of insecticides such as iodofenphos, fenitrothion, arprocarb or diazinon, sprayed liberally into infested floor cavities, pipe runs, areas behind sinks, pipes and stoves may deal with a localised infestation. For tiled surfaces an oil-based insecticidal spray or an insecticidal lacquer may provide long-lasting protection, but these are best applied by a pest control company. Boric acid powder and baits are also used by such specialists, with considerable success. Several treatments are usually necessary to eliminate a cockroach infestation and commercial premises are best protected by a pest control contract providing regular inspections.

Related Species

The large American Cockroach *Periplaneta americana* which can grow to 44 mm ($1\frac{3}{4}$ in) and the Brown Banded Cockroach are occasionally introduced in ships, baggage or containers and become established in premises such as heated animal houses in zoos or the heating systems of large buildings. Other indoor species occur in Australasia, the USA, South Africa and Japan.

Biscuit Beetle

Identification

The Biscuit Beetle *Stegobium paniceum* is a small reddish-brown insect which belongs to the same family as the Common Furniture Beetle which it closely resembles.

Only about 3 mm ($\frac{1}{8}$ in) long, it is one of the 'stored product insects' or 'pantry pests' that may infest food in domestic larders or in other storage areas.

In the adult the head is completely concealed from view under the rounded hood or 'pronotum' which projects forwards from the thorax.

Life Cycle

About 100 eggs are laid separately over a period of three weeks in the foodstuff or nearby crevices. The larval stage takes 4 to 5 months,

97

including four moults and the newly emerged larva is very small and active. At this stage it can survive starvation for a week. The full grown larva is about 5 mm (about ¼ in) long and white in colour. It turns into a pupa surrounded by a cocoon of food particles and this stage lasts from 12 to 18 days.

Damage Caused

Biscuit beetles are attracted by all types of flour goods, cereals and items such as soup powders and certain spices.

The species has even been found thriving on such poisonous substances as strychnine, belladonna and aconite, hence its American name of Drug Store Beetle.

It has also been found among books and manuscripts and has been recorded as boring straight through a shelf full of books and penetrating tin foil and sheet lead.

Control Measures

Thorough cleaning out of any food residues infested by Biscuit Beetle and the use of a household product sold for crawling insects, such as a puffer pack of carbaryl insect powder or an aerosol containing diazinon will deal with this insect at any stage of its seven month life cycle.

Related Species

The Common Furniture Beetle is dealt with under 'woodworm'.

Booklouse

There are several species of Booklice, also known as *Psocids*, and they are classed in a separate group (the *Psocoptera*) from the parasitic lice.

Almost all the indoor species are wingless and no males are produced, parthenogenetic reproduction being the rule.

All of them run quite rapidly and this, despite their very small size often attracts the householder's attention. One species, *Trogium pulsatorium*, produces an audible tapping noise with its abdomen against paper or wood and it has been suggested that this species may have accounted for some references to Death Watch Beetle.

Identification

The typical Booklouse *Liposcelis divinatorius* is very small, only 1 mm ($\frac{1}{32}$ in) long and with a soft cream or light brown body and prominent but thin antennae.

Life Cycle

The sticky pearl-coloured eggs, each almost a third the size of the adult, are laid singly and become cemented to the food material. There is no larval stage, the four nymphal stages resembling the adult.

Temperature and humidity affect the length of life cycle but in laboratory conditions 200 eggs are laid, they hatch in 11 days, the nymphs mature in 15 days and adults have lived for six months.

Damage Caused

Booklice feed on microscopic moulds that grow on the glue of book bindings or on damp cereal food, pasta, or the surfaces of plaster, leather, books or woodwork.

As the moulds only grow in moist conditions the lice are found in ill-ventilated damp areas or sometimes new homes in which the plaster is still drying out.

Control Measures

Thorough airing and drying of an infested area, plus removal of all signs of mould, will get rid of booklice. They cannot live in dry cold or dry

Confused Flour Beetle.

heat. A proprietary insecticide aerosol or powder for crawling insects should be used where they persist. Destroy badly infested food.

Related Species

The various parasitic lice are covered under 'Lice'.

Flour Beetles

Small reddish-brown beetles about 3–4 mm (about ⅛ in) long that feed on flour and cereal debris in warm buildings. May be accidentally introduced into the larder in packaging or in the ingredients themselves.

Commonest species are the Rust Red Flour Beetle and the Confused Flour Beetle *Tribolium Confusum* (which in turn is often confused with the Rust Red Flour Beetle). Flour beetles may produce five generations in a year, and adults can live for over a year. The eggs stick to flour particles and the yellow-brown larvae, about 6 mm ($\frac{1}{4}$ in) long, crawl about very actively.

Control Measures

Clear out infested food, clean the area thoroughly and apply a long-lasting spray insecticide described as suitable for crawling insects. Infested mills, warehouses and cargoes need professional fumigation with methyl bromide gas.

Flour Moths

Moths whose grubs feed on stored food—especially cereals, chocolate, cocoa, dried fruit, nuts and any sort of flour product.

Adults are mottled grey and brown, 10–15 mm (about $\frac{1}{2}$ in) long. Larvae are dirty white with brown heads and grow up to 12–15 mm (about $\frac{1}{2}$ in) long, spinning a light matted webbing as they feed. Both grubs and adults therefore are pests as guests in the kitchen.

Control Measures

Destroy infested food, or return it to the shop with a strong complaint. Clean out the store cupboard and spray surfaces with a residual insecticide for crawling insects. Infestation in commercial premises needs professional treatment by pest control contractors.

Grain Weevil

Occasionally gets from flour mill to larder. Characteristic long 'snout' for boring into grains of cereal. A major pest of grain silos, *Sitophilus granarius*, is a problem on many farms.

Grain damaged by weevil.

Infested grain in bulk can be treated with malathion, iodofenphos or fumigated with aluminium phosphide tablets or methyl bromide gas.

Related Species

Grain beetles such as the Merchant Grain Beetle and the Rust Red Grain Beetle about 4 mm ($\frac{1}{8}$ in) long infest whole grain products but lack the weevil's distinctive elongated head. Recent excavations have shown that some grain beetles and flour beetles have been in Britain since Roman times, including the Saw-Toothed Grain Beetle and the Flat Grain Beetle.

Larder Beetles

One of the larger beetles that may occur in homes, often associated with a dead rodent or dead bird in some inaccessible part of the roof or under the floor.

Also known as the Bacon Beetle it used to attack bacon, ham and cheese when home curing was customary.

Identification

The adult Larder Beetle *Dermestes lardarius* is a large oval beetle about 8 mm ($\frac{1}{4}$ in) long, with a pale grey band across the front half of its wing cases. It is one of four species also known as Hide Beetles.

The dark coloured grubs are fairly conspicuous because of their size, 10–15 mm (about $\frac{1}{2}$ in) long when fully grown.

Life Cycle

Eggs are laid over a period of several months among the larval food supply. The larvae are dark coloured with tufts of hairs and the skins are moulted six or more times as the larvae grow.

When seeking a site to pupate, the larva will bore into materials, including wood. The life cycle takes two or three months and the adults may hibernate in unheated buildings.

Damage Caused

All four dermestid beetles are scavengers (the family name literally means 'Skin-eaters'). Larvae infesting birds' nests, dead animals or birds, may invade food cupboards.

They can be a serious pest in hide warehouses, tanneries and premises where animal products are processed.

Control Measures

Clean out all possible reservoirs of infestation such as old birds' nests, scraps of fur or damaged leather. Treat the area with an insecticidal powder or aerosol containing diazinon or lindane or other preparation designated for crawling insects.

Related Species

Dermestes haemorrhoidalis and *Dermestes peruvianus*, neither of which has a common English name, appear to be becoming more widely distributed than the Larder Beetle which they resemble but lack the grey band across the wing cases. The fourth species, *Dermestes maculatus*, most often called the Hide Beetle, has some white scales on its otherwise dark wing cases.

Mealworm Beetle

Identification

The Mealworm *Terebrio molitor*, is a dark, almost black beetle between 12 and 16 mm ($\frac{1}{2}$ – $\frac{5}{8}$ in) long with glossy wing cases.

The larvae are long, up to 28 mm ($1\frac{1}{8}$ in), clearly segmented and bright yellow with a waxy appearance.

Life Cycle

About 500 bean-shaped eggs are laid in food debris and the larvae (mealworms) grow by a series of moults until they turn into curved pupae.

The larval stage lasts about six months and the adults live about three months.

Damage Caused

Mealworms are found in old buildings where grain or flour products have been stored and the residues become driven into the crevices or spaces in the structure. They prefer damp, dark areas and feed on farinaceous food and such debris as dead insects.

Control Measures

Find and clear up the source of infestation. Rigorous brushing and vacuum cleaning of all gaps between boards and skirtings will help. Make liberal use of a household crawling insect powder based on lindane or carbaryl puffed into the areas from which adult beetles emerge or where mealworms are seen.

Plaster Beetles

In a new house, or where damp plaster occurs in old property, very small dark-coloured beetles may be encountered, only about 2 mm ($\frac{1}{16}$ in) long. These insects and their larvae feed on the moulds and mildews which grow on damp walls. If the rooms are heated and thoroughly ventilated the moulds will die and so will the beetles.

Identification

There are six species of beetle known as Plaster Beetle or Fungus Beetle. The three fungus beetles are all of the genus *Cryptophagus*, the commonest in houses being *C. acutangulus*. A typical plaster beetle is *Lathridius minutus*.

Because of their very small size expert examination with a lens is necessary to determine the species.

Life Cycle

Eggs are laid singly, hatch within a week, the larval stage lasts about two weeks and the complete egg-to-adult cycle takes from 20 to 30 days depending on humidity and temperature.

Damage Caused

As these beetles only feed on moulds and mildew they cause little damage but may carry mould spores from place to place and between food commodities.

Control Measures

Dry heat and thorough ventilation of premises plus disposal of any infested food will get rid of the mould and the beetles will die out. Occasional heavy infestations in warehouses or packaging may need fumigation. The persistent presence of fungus beetles may be a symptom of a condensation problem and advice should be sought on its cure.

Spider Beetles

The Australian Spider Beetle and a related species, the Golden Spider Beetle, are so-called because they resemble small, round spiders. They are often found in stored products in food warehouses and may be transferred in packets of stored foods to domestic larders where they feed on any food debris in cracks and crevices. Birds' nests may also be reservoirs of infestation.

Identification

The Australian Spider Beetle, *Ptinus tectus*, is dull brown with a prominent abdomen and long legs. The Golden Spider Beetle, *Niptus hololeucus* is similar but is golden yellow with soft hairs on the wing cases. Both species are from 3–4 mm (about ⅛ in) long. Adults may feign death when disturbed.

Life Cycle

About 100 sticky eggs are laid singly or in batches and the white fleshy grubs are covered with fine hairs. They roll up when disturbed and when ready to pupate the grubs wander about sometimes biting through cardboard or wood before forming a brown cocoon.

The total life cycle takes three to six months.

Damage Caused

Spider beetles are basically scavengers and infestations often originate in old birds' nests from which the insects drop or crawl into rooms. They feed readily on cereals, spices and a wide range of debris that may accumulate in neglected larders or store-rooms or in floor cracks under skirting boards.

A residual insecticidal powder dusted well into cracks and crevices from a puffer pack will control any adult spider beetles and kill any larvae that subsequently emerge from eggs.

Check that no old cereal packet or old nest is acting as a breeding place.

Silverfish

This primitive, wingless insect often appears in baths or other damp situations. It does not crawl up the waste pipe as many people think, but falls into the bath probably while seeking water, and is unable to escape again.

Silverfish.

Identification

The Silverfish *Lepisma saccharina* is silver-grey with a 7 mm ($\frac{1}{4}$ in) long cigar-shaped body and three 'bristles' at the end of its abdomen. It runs very quickly when disturbed or suddenly exposed to light.

Life Cycle

Eggs are laid in tiny cracks and crevices and hatch in three to four weeks producing pale whitish nymphs which mature through a series of moults. The adult can live for three years or more in warm, moist surroundings.

Damage Caused

Silverfish do little or no harm apart from minor damage to paper, book bindings or damp wallpaper. They feed mainly on minute residues of carbohydrate material such as starch, wallpaper paste and possibly mould growth. Their presence may indicate some defect causing unnecessary dampness.

Control Measures

Check the source of any undue dampness, such as faulty plumbing, rising damp or condensation that provides the moisture favoured by silverfish.

If they persist, a puffer pack of carbaryl insect powder may be used to treat cracks or crevices around plumbing fittings, or other such areas in which silverfish occur. An insecticidal aerosol will give similar protection in liquid form.

Related Species

The Firebrat *Thermobia domestica* is similar in overall appearance to the silverfish, but greyish white instead of metallic looking and slightly larger. The firebrat likes warm, dry conditions and sometimes occurs around boilers, hearths or ovens.

Textile Pests

Carpet Beetles

People are often perturbed or puzzled to find small golden-brown 'woolly bear' grubs in an airing cupboard, at the edges of carpets or in a piece of stored felt. These are the larvae of one of the species of carpet beetles which thrive in warm, dry conditions. They seem to be increasing to some extent in premises with central heating and fitted carpets. The adults often turn up during spring cleaning and the larvae in Autumn when they wander about seeking food and hibernation sites.

Identification

The adult Varied Carpet Beetle *Anthrenus verbasci* is small, oval in shape, not unlike a small mottled brown-and-cream ladybird. Size of adult varies from 2–4 mm ($\frac{1}{12}$ –$\frac{1}{8}$ in), the larvae are about 4 mm ($\frac{1}{8}$ in) when full grown.

Life Cycle

Outdoors adults feed on nectar and pollen of light coloured flowers and, after mating, the female often enters houses and each lays between 20 and 100 eggs in birds' nest or accumulated debris in roof voids, under the eaves or in carpet fluff.

The larvae feed on feathers, dead insects, fur and wool, having the ability, which they share with clothes moths, to digest keratin. From such situations as lofts and eaves they wander along pipe-lagging and into upper rooms, airing cupboards, wardrobes or on to carpets, furs or upholstery.

Eggs are laid in Spring and hatch in about ten days. The life cycle usually takes about twelve months, the larvae moulting as they grow and the dry cast-off skins are often the first sign of infestation. Adults appear in Spring and early Summer and most of the larvae spend the Winter in hibernation. The grubs can survive starvation for periods of up to nine months.

Damage Caused

The damage caused by Carpet Beetles usually consists of well-defined round holes often along the seams of fabric where the grubs bite through the thread. In Britain, the Carpet Beetles have now outstripped the Clothes Moth as the major textile pests, damaging all types of woollens. They are well established elsewhere in Europe and in the USA.

Control Measures

If Carpet Beetles are found, the first task is to trace their origin. Search the loft and eaves for birds' nests or the dead bodies of birds or rodents and remove them. The link between insect infestation and birds' nests is well established. Check pipe lagging and vacuum clean shelves, floorboards, cupboards, carpets and upholstery.

Spray affected carpets or furnishings with a proprietary Mothproofer aerosol which contains lindane. A persistent contact insecticide of this type is necessary because the 'woolly bear' larvae often survive exposure to a shorter-lived insecticide. Dust between floor-boards, under carpets and crevices with a proprietary insect powder based on carbaryl or lindane.

In a confined space, to protect clothes in a trunk or case, crystals of moth repellent paradichlorobenzene or naphthalene will keep out carpet and fur beetles. Use in accordance with the label and ventilate thoroughly to disperse the odour before unpacking for use. Avoid direct contact with delicate fabrics.

Fur Beetle.

Related Species

The Varied Carpet Beetle is the commonest species in the South of England, but the more widespread Fur Beetle *Attagenus pellio*, which is black with a white spot on each wing case, and the rarer Black Carpet Beetle have similar habits and all may damage woollen fabrics, furs and hair. The Museum Beetle damages mounted specimens of other insects as well as textile materials.

Clothes Moth

The Common Clothes Moth and the Brown House Moth can conveniently be dealt with together as the commonest moths affecting stored clothing in Britain. The Case-Bearing Clothes Moth, White Shouldered House Moth and Tapestry Moth also share the wool-eating habit and may be encountered indoors.

Identification

The true clothes moths fold their wings tent-wise, forming an angle over the abdomen while the house moths fold their wings almost flat along their bodies. The wings of the Common Clothes Moth *Tineola bisselliella* are a uniform pale buff and a fringe of hair marks the borders of each wing. The Brown House Moth *Hofmannophila pseudospretella* has flecked golden-bronze wings. Both species tend to run rather than fly. Most 'clothes moths' are in fact Brown House Moths and this species is probably found at some time in every home in Britain. Adults are about 7.5–10mm (just under $\frac{1}{2}$ in) long.

Life Cycle

The adult female lays between 50 and 100 eggs which are attached to the fabric on which the larvae are to feed. The grubs, which hatch in about ten days, are small creamy-white caterpillars with brown heads, and those of the clothes moths conceal themselves under a loose web-like gallery of silk which they secrete. The Clothes Moth larva grows to 10 mm (nearly $\frac{1}{2}$ in) long, that of the House Moth up to 18 mm ($\frac{3}{4}$ in). The larvae take from three to eighteen months to develop and the pupal stage lasts from 10 to 60 days within a silken cocoon in the case of the House Moth. The adult moth lives for up to four weeks. The Case-Bearing Clothes Moth produces a cylindrical open-ended case of this silk and attaches to it fibres of the material on which it feeds, making detection more difficult.

Damage Caused

Adult Clothes Moths and House Moths do not eat any material and in fact are incapable of feeding at all. All the feeding and, therefore, the damage is done by the larvae and stops as soon as they pupate.

Clothes Moths and House Moths seldom develop in absolutely clean wool and fur, and prefer material that has been dyed or soiled by food, perspiration or urine.

The Brown House Moth also feeds on many commodities including leather, dried fruit and cork. It can survive long periods of unfavourable conditions by going into a state of suspended animation known as a 'diapause'.

Control Measures

Preventative measures are best in order to avoid moth damage. Keep clothes scrupulously clean and store woollens in sealed polythene bags or closely wrapped in paper, in tightly closed drawers or cupboards preferably in a cool room. Before putting woollens away, fold in one or two discs or tablets based on paradichlorobenzene or naphthalene which have a fumigant effect. Hang moth repellents or dichlorvos vapourising strips in wardrobes and cupboards.

White Shouldered House Moth.

Clean carpets regularly and every six months spray the double thickness at the edges of fitted carpets and areas under heavy furniture with a residual insecticide such as that contained in proprietary mothproofer aerosols. Spray both surfaces at the edge of a carpet and into the cracks between floorboards. Upholstery can also be protected by placing the repellents down the sides and backs—the latest proprietary forms do not have the pungent smell of 'moth balls'—but where moth damage has already occurred it is necessary to remove the covering and spray with mothproofer.

Related Species

The Case Bearing Clothes Moth has three dark spots on its buff wings and appears duller than the Common Clothes Moth. The larva drags itself about in a cylindrical open ended case made of cut off fibres and wool. The White Shouldered House Moth has a white patch where the wings join the thorax and a white head.

Mites and Spiders

Mites

Mites are not true insects. They often escape detection by the householder until their numbers build up to a serious infestation. The exception is the bright red Clover Mite (or Red Spider Mite) *Bryobia praetiosa*, which is relatively conspicuous.

Identification

Mites are characterised by their small size, almost on the limit of vision by the naked eye, the possession of eight legs and a bag-like unsegmented body.

Unlike spiders, which also have eight legs, mites have a larval stage in which they have only six legs.

The identification of individual mite species can often only be made with the aid of a microscope and most species infesting food or furniture in buildings are described just as *Tyroglyphid* mites.

Life Cycle

The typical mite starts life as an egg laid in a crevice and in a few days a six legged larval stage emerges, feeds for a few days and then grows by moulting until it becomes an adult. Some species can overcome adverse conditions for as long as six months by assuming a cyst-like form called a hypopus.

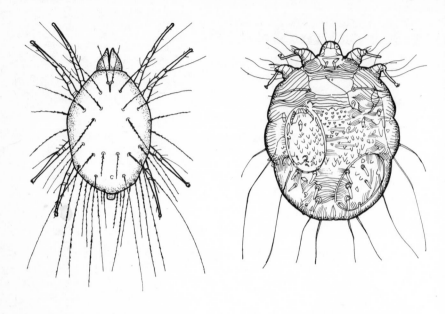

House/Furniture Mite.   Itch Mite.

Damage Caused

The minute House Mite, or Furniture Mite, *Glycyphagus domesticus* has been described as 'Living Dust'. It infests foodstuffs and may be present in upholstered furniture in damp, ill ventilated rooms. The larder may be invaded by the Flour Mite *Acarus siro* which is a common pest of flour, cereals, cheese and dried fruits. The tiny House Dust Mite *Dermatophagoides pteronyssinus* is probably present in every bed mattress where it feeds on microscopic particles of human skin, and in a few people it causes an allergic reaction with symptoms of asthma.

Two species of mite which cause unpleasant effects on man are the Harvest Mite and the Itch or Scabies Mite *Sarcoptes scabei*. The larval stage of the Harvest Mite is accidentally brought indoors on dogs, cats or on outer clothing to which it transfers from long grass or shrubs. It is common in some grassland areas and causes severe irritation when it bites humans.

116

Mange in domestic animals is caused by mites and the disease of scabies in humans is caused by the Itch Mite which burrows into the skin causing severe irritation and a skin rash. 'Grocers Itch' is an allergic reaction to the Flour Mite and this mite imparts a bitter or 'minty' taint to food that it infests.

Bird Mites sometimes enter buildings from the old nests of starlings, sparrows, house martins or poultry and may cause irritating bites.

Occasional invasions of the large numbers of the bright red, Gooseberry Red Spider Mite or Clover Mite usually originate in the garden. The females migrate into houses in Spring, seeking egg-laying sites and in Autumn they seek sites in which to hibernate.

Control Measures

Treatment consists of disposing of affected food, ventilating and drying the larder, or, heating the room to a high temperature for 24 hours. Flour mites die at about 35°C (95°F). Follow up by spraying surfaces and crevices with an insecticidal aerosol.

The use of repellents such as dimethylpthalate or undiluted benzyl benzoate will help to prevent mites attacking the skin and these should be applied to cuffs and socks. Several proprietary insect repellent creams are based on these. An emulsion of benzyl benzoate applied to the skin will relieve scabies. An emulsion of 1 percent malathion or lindane around doors and windows will usually control Clover Mites and it may be necessary to cut back vegetation for about 3 ft (1 m) from the walls of the house.

Ear Mites in cats or dogs need prompt veterinary attention.

Spiders

Spiders feature in many legends, from Ariadne and Robert the Bruce to Miss Muffet.

In Britain they cause no harm, but most people regard them with feelings of revulsion. The old English name was 'Attercop' meaning Poison Head. They feed on other insects and are, therefore, beneficial. Spiders, having eight legs, are not insects but are Arachnids. Three species are common in houses and are responsible for the household cobwebs. In Australia, the Americas and the Tropics, certain poisonous spiders come into houses and control measures are then desirable and necessary.

Identification

The general form of spiders is familiar to all householders. There are said to be three spiders in each room of every house no matter how house-proud the owner.

They have a rather globular body, eight long legs and a pair of 'palps' looking like extended jaws on the head instead of antennae. In male spiders the palps are sex organs. Their body has only two main divisions, the cephalothorax and abdomen and they have four to eight simple eyes.

Not all spiders produce webs, but the House Spider's presence is most often detected by the untidy 'cobweb' with which it ensnares small insects.

The House Spider *Tegenaria domestica* and the larger dark brown *Tegenaria atrica* are the species most often noticed indoors. The large garden spider *Aranea diadema* is the one most often responsible for the beautiful web out of doors in Britain. *Ciniflo similis* is one of the commonest spiders in cellars and sheds.

Life Cycle

Spider eggs are surrounded by a protective covering of silk spun by the adult female. Several hundred eggs are laid and the first spiderlings to hatch and moult eat many of those hatching later. The young spiderlings disperse, often on strands of silk 'gossamer' that are caught by the wind, and they are miniature versions of the adult.

Damage Caused

Spiders although beneficial in destroying insects in the house, cause feelings of revulsion in many people, to the extent that some will not go into a room where they know there is a spider.

The typical spider in the bath has not come up the waste-pipe but has fallen into the bath from the overflow or the bath surround and is unable to climb the smooth sides. In the USA, the Black Widow, the Brown Recluse and in Australia, the Red-Back and the Sydney Funnel Web can inflict serious and sometimes fatal bites.

Control Measures

If you wish to remove a spider without killing it or washing it away, place a small carton inverted over it and slide a piece of card or stiff paper between the opening and the surface on which the spider rests.

Any household insecticidal spray or aerosol will kill spiders and scrupulous cleaning will discourage them.

Related Species

Harvestmen are like extreme spiders, they have eight exceptionally long thin legs but they do not spin webs and only occasionally enter homes.

Woodboring Insects

'The wormes do brede in soft woode and sweete'.   Elyot

Termites

Termites occur throughout the world from approximately 40° North to 40° South and two species occur in Europe as far north as Paris.

They are the most destructive of the timber pests and are highly organised social insects. There are two main types, the Drywood Termites and the Subterranean Termites making a total of about 2,000 species.

Identification

Termites are also known as 'White Ants', but are neither white, nor are they ants. They are far more ancient than the ants and mostly brown in colour, with paler patches although the workers resemble large ants in general appearance and in their community habits.

Life Cycle

Termites hatch from whitish eggs and have no distinct larval stage, but the immature termites resemble the adult workers.

There is a complex caste system, at the centre of which is the queen which may lay several eggs a minute for 25 years and have an abdomen 10 cm (4 in) long. Some queen termites live for 50 years.

At certain times of the year, usually in warm humid weather, the winged reproductive termites are produced and swarm into the air,

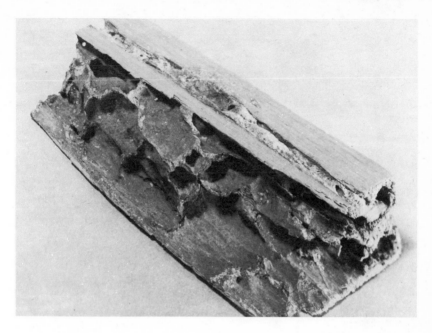

Termite damage.

usually at dusk.

In the Drywood species, the fertilised queen will seek egg laying sites in old timber or a wooden building. The Subterranean queen will seek a crack in the earth in which to start a new colony.

The termites will develop into workers with various tasks to perform, or soldiers to defend the nest from predators, or 'nursemaids' to tend the young and the queen.

The Drywood termites live on wood and digest cellulose directly. The Subterranean species gather cellulose material and use it to grow 'fungus gardens' on which they feed.

Damage Caused

The destruction of wood and other cellulose based materials by termites

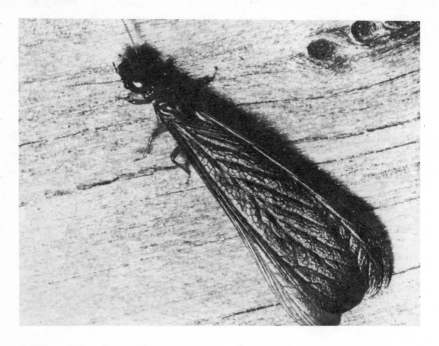

A Winged Termite or Alate.

costs hundreds of millions of pounds a year. Structural timber and cladding is eaten away from inside and the foraging Subterranean workers will build mud shelter tubes from the ground up into the structure in which they are feeding. Some species damage crops and growing trees.

Control Measures

New building sites can be protected by treating the soil under and around the concrete base by a special termite fluid containing a residual insecticide such as dieldrin.

New timber should be treated by vacuum-pressure impregnation with a CCA type wood preservative.

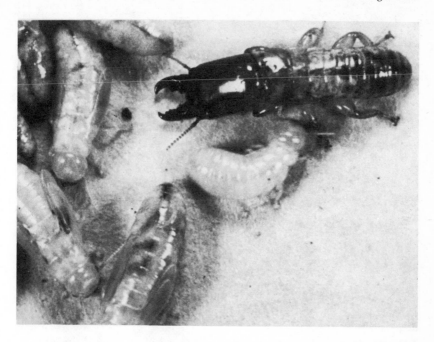

A Soldier Termite.

For existing buildings it may be possible to spray a termite killing fluid over all the timber but it is difficult to reach all the vulnerable areas. In such cases 'tent fumigation' is the answer, in which gas proof sheets are rigged right over the house and weighted down. The building is then filled with a gas such as Vikane (sulphuryl fluoride) which penetrates the wood, and subsequently ventilated.

Woodworm

In times past, the presence of woodworm holes was supposed to add authenticity to a claim that an article was antique and unscrupulous dealers were said to drill 'woodworm holes' in reproduction furniture.

The Common Furniture Beetle *Anobium punctatum*—known as the

Here, a table has been severely attacked by wood-boring beetles which have reduced the inside of the wood to a mass of sawdust.

House Borer in the USA and New Zealand—certainly does not confine its attention to old furniture. It is also the chief form of woodworm in softwood structural timbers and attacks quite new plywood, especially birch, and wickerwork. Surveys indicate that it is present in almost three out of every four houses. The Common Furniture Beetle is the species you are most likely inadvertently to pass on the stairs as it accounts for over three quarters of all the forms of woodworm infestation in Britain.

Identification

The first signs of woodworm are the neat round exit holes about 2 mm ($\frac{1}{12}$ in) across that appear in wooden surfaces, sometimes accompanied by

The larva of the Common Furniture Beetle spends usually three years boring through the timber in which it was deposited as an egg.

small piles of white wood dust beneath the holes. Fresh holes show clean white wood inside. Old holes are dark with dirt and dust.

The holes are made by the emerging adult beetle. The commonest, the Common Furniture Beetle, *Anobium punctatum* is a small brown beetle which can be from 3 to 6 mm ($\frac{1}{8}$ – $\frac{1}{4}$ in) long, its head concealed by the hoodlike thorax. Under a lens the wing-cases will be seen to be pitted with rows of tiny depressions. The beetle flies quite readily and in the Summer months often shows up on light coloured paintwork around windows.

Life Cycle

The eggs are laid in groups of two or three in cracks or joints in rough unpainted wood or in old flight holes. The emerging larva 'the woodworm' bites its way straight into the wood leaving no visible trace of its entry and it spends up to three years tunnelling in the wood making galleries at a rate of about 5 cm (2 in) each year. Wood dust and faecal pellets make up the bore-dust or 'frass' which packs the tunnel behind the grub and may fall out of a few flight holes.

A pupal chamber is constructed just under the surface of the wood and the adult beetle bites its way out with a circular motion, producing the perfectly round hole.

Adults emerge during the late Spring to late Summer.

Damage Caused

Out of doors, the Furniture Beetle's natural habitat is in dead branches or damaged trees but in buildings it may seriously weaken structural timbers, flooring or joinery and can damage furniture so as to reduce seriously its value and usefulness.

In time, it could literally eat us out of house and home. Fortunately, it is a slow eater.

The Furniture Beetle may fly into a house and land on the rafters or joists or it may be introduced in old packing cases, tea chests, wicker baskets, or secondhand furniture of the type that is 'Queen Anne front and Mary Ann back'.

Hickin estimates that remedial work for woodworm attack in Britain costs over £15 million ($30M) each year—a figure that increases each year.

Furniture of Cuban mahogany or teak is immune from attack, but almost all other wood used for furniture or house construction is susceptible, especially the sapwood which comprises 50 percent of modern softwood timbers. The smaller cross-section dimensions of rafters and joists in new housing indicate that the woodworm problem will become more serious unless such timbers are preservatively pre-treated. Vacuum pressure impregnation of timbers is available from many timber merchants for this purpose.

Other wood-boring beetles (see 'Related Species') attack specific types of wood or are confined in geographic range but contribute to the total damage.

Control Measures for Woodworm in Furniture

Old-fashioned remedies such as wiping furniture with turpentine do not give lasting protection and there are on the market very effective woodworm killing fluids which consist of a special insecticide in a very light and penetrating solvent. These will deal permanently with all stages of the woodworm life cycle in one thorough treatment. However, it is not sufficient only to treat areas where holes are seen, as grubs may be tunnelling anywhere in an infested item. All surfaces should be thoroughly coated by brush or spray, including inside drawers, backing, undersides and feet.

In addition, extra penetration into furniture or joinery may be obtained by using a special injector with a patent nozzle to squirt some of the woodworm killer into the holes about every four inches (10 cm). Small plastic injector bottles or aerosols with special tubes attached are available for this. In practice it is best to inject the holes before coating a surface so that you can see which you have done and which not. Eye droppers and hypodermic needles are not suitable for injecting the fluid!

The best brands of woodworm killer should not harm polished or varnished surfaces but over the years an old piece of furniture may have had so many surface layers of various materials applied that it is wise to test fluid on a small inconspicuous area at first.

Valuable items, especially if upholstered, may cause the owner to feel apprehensive about spraying them. In this case they can be fumigated in specially constructed chambers with a gas that will kill all stages of woodworm, moths or any other live-stock.

As a precaution against woodworm and to prevent reinfestation of fumigated items it is sound practice to use a special insecticidal polish. Such polishes should also be applied occasionally to unpolished parts of untreated furniture to prevent adult beetles laying eggs there.

Control Measures for Woodworm in Structural Timbers

Woodworm in furniture should never be neglected. It may spread to the rafters and the floor and the higher proportion of sapwood in today's softwood building timbers make them readily susceptible to an attack by woodworm—attack which they are less able to withstand because of the

127

smaller dimensions of modern rafters and joists. Roofs, floor joists and staircases are often sites of woodworm infestation.

Treatment of structural timbers affected by woodworm is best carried out by one of the reputable woodworm and dry rot control specialists in membership of the British Wood Preserving Association. The best of these offer free surveys, detailed reports and estimates and guarantee their workmanship for twenty or even thirty years. This written guarantee, if from a reputable company, is valuable evidence of treatment when the home changes hands.

If you wish to tackle an infestation of woodworm yourself the appropriate brand of woodworm killing fluid is available from builder's merchants and Do-it-Yourself shops in drums. If you decide to deal with woodworm yourself, remember, it's no use only treating the area where you see woodworm holes—other grubs may be active but unseen in the next piece of timber.

You can treat roofs against woodworm yourself like this, but it is often wiser to call in a specialist company.

Thoroughness is the key-note to success in all timber treatment work. For woodworm attack in rafters, joists and flooring apply woodworm killer with a coarse spray at the rate recommended on the can after thoroughly cleaning down the surfaces with brush and vacuum cleaner.

To estimate the timber area in a boarded roof, take the overall dimensions, and add twice the length of a rafter multiplied by its depth and by the number of rafters. If the roof is not boarded, add the width of a rafter to twice its length, multiply by its depth and by the number of rafters. Use a similar procedure for the joists and purlins.

If the timber is very dry it may soak up more fluid so estimate on the generous side.

There is no best time of year for treatment. Modern woodworm killing fluids are very efficient, will kill the insects at all stages of their life cycle and prevent future attack, provided all surfaces of timber are treated.

Any roof insulation between joists must be removed before treatment and only replaced after the treated wood has dried.

Cover the cold water storage tank if it is in the roof to protect it from fall-out. If the tank is lagged with expanded polystyrene insulation, don't let the fluid come into contact with it or your insulation may dissolve. Any rubber-covered cables exposed should be painted with a wood sealer before you start spraying, as the fluid is harmful to rubber. Make sure any electrical wiring in the area is sound and well insulated. Do not smoke while spraying.

Wear a pair of old leather, rather than rubber, gloves unless you are prepared to sacrifice a cheap rubber pair. Other sensible precautions are to rub a barrier cream over exposed skin on face and wrists and to wear a light fume mask to avoid inhaling the vapour in the confined space of a roof. Goggles will protect the eyes if they smart.

Selection of a sprayer is important but some garden sprayers are suitable if they can be pumped up well and maintain good pressure. Ideally it should hold a gallon at a time and have a fairly coarse nozzle producing a fan spray pattern. Too coarse a nozzle may mean that you drench the joists too much and the run-off stains the ceiling below. Too fine a nozzle will vaporise the spray and make the work unpleasant and not very effective. An extension lance is essential for reaching into the apex, eaves and less accessible areas. As you spray, the fine fall-out of fluid will be sufficient to protect the laths between the joists.

If woodworm occurs in floorboards or floor joists, take up every fourth

The Death Watch Beetle
adult is similar in shape,
but twice the size, of the
Furniture Beetle at 6 mm
($\frac{1}{4}$ in).

or fifth board so that you can reach between the joists and the undersides of the boards. Cover the floor with a large sheet of polythene if you wish to replace carpets, otherwise wait a week for the surface to dry out. Should you accidentally stain plaster with woodworm fluid, leave it for a week or two and if it still shows, coat it with an aluminium primer before redecorating.

Rubber or plastic floor covering, either tiles or in sheet form should not be laid over wooden flooring that has been treated with timber fluids.

No new house need ever suffer from woodworm or rot. Modern vacuum-pressure impregnation processes make protected timbers widely available at low cost from timber merchants who operate the treatment plants.

The extra cost of pre-treating all the structural timber for an average house is only a very small sum when one considers what a severe attack of woodworm in an unprotected property may cost to cure.

If there is no apparent damage at the time of a timber survey, it is now possible to insure all the timber in a house against damage by woodworm or rot.

Wood Boring Weevil.

House Longhorn Beetle.

Related Species

Best known, or most notorious of Britain's other wood-borers is the Death Watch Beetle *Xestobium rufovillosum*. As it does not fly indoors, this species is becoming rarer in buildings as the massive old hardwood beams that it inhabits are either destroyed or treated. Most Death Watch infestations originated in over-mature timber used when the buildings were first erected. It is still common in dead wood out of doors.

Death Watch attacks in buildings decrease northwards and are unknown in Scotland. A larger beetle than *Anobium*, it leaves correspondingly larger holes, about 4 mm ($\frac{1}{8}$ in) across and has a life cycle that takes up to ten years to complete.

The rapid tapping sound that gives the Death Watch its name is made by striking the head against the surface on which it stands and is produced by both sexes.

The Death Watch rarely attacks softwoods.

The Wood Boring Weevils, always closely associated with wet rot, are increasing in south east England and the Powder Post Beetle sometimes emerges from hardwood floors or panelling. The female Powder Post Beetle only lays eggs on wide-pored hardwoods of 3 percent starch content. The larger House Longhorn Beetle, a major pest of softwood on the Continent and in South Africa, is chiefly confined to areas of north west Surrey.

Two species of Wood-boring weevils, *Euophryum confine* and *Pentarthrum huttoni* are involved in boring into wood that has already been attacked by wet rot fungus. Both have the typical long snout of weevils and both adult and larva tunnel into wood, producing ragged or roughly oval holes. Adult weevils are said to live for well over a year. *Euophrym* was not recorded in Britain until 1937 but is now well established in London and the Home Counties.

The House Longhorn *Hylotrupes bajulus*, is the largest British woodborer, an adult female being up to 1 in (25 mm) long. It is dark brown with two greyish marks on the wing cases and long antennae. The white grub may reach a length of 24 mm and feeds voraciously on softwood roof timbers. The large flight holes are oval and there may be only a few visible in a veneer of sound wood concealing extensive damage beneath. The life cycle takes from three to six years or longer. It is called the Old House Borer in South Africa and the USA.

The Wharf Borer *Nacerdes melanura*, a large brown beetle with prominent antennae, occasionally emerges from wet timbers around riversides or from damp basements into buildings.

Control measures against all these beetles are similar to those for Furniture Beetle but damaged wood must first be scraped or cut away. If any of these species is suspected it is wise to call in expert advice.

The Bark Beetle *Ernobius mollis* resembles *Anobium* but is confined to the bark attached to timber and will die out if this is removed.

Rots and Moulds

Dry Rot and Wet Rot

Dry Rot and Wet Rot are forms of fungal decay often due to faulty construction or poor maintenance. They thrive on damp wood in conditions of poor ventilation, and can completely break down timber unless checked and eradicated. First signs may be warping or cracks in skirting or window frames and a penknife will easily sink into rotting wood. If neglected Dry Rot can spread under favourable conditions at a rate of 12 ft (3·6 m) a year and recent research shows that it will grow in relatively cold conditions.

Identification

True Dry Rot is the fungus *Serpula lacrymans* and it requires over 20 percent moisture content in the wood to germinate from minute airborne spores. It typically causes cuboidal cracking in the timber and grows by the many small thread-like hyphae which penetrate wood, brick and mortar. It produces specialised hyphal strands called rhizomorphs which carry moisture as the growth breaks down the cellulose in the wood. Wet Rot requires about 40 percent minimum moisture content. Advanced attacks of Dry Rot show matted whitish strands of fungus, large plate-like fruiting bodies (sporophores) and rust red spore dust. The commonest form of Wet Rot (*Coniophora puteana*) often leaves a thin veneer of wood over a dark, brittle mass. The hyphae are dark, often forming a fern-like pattern and the olive brown sporophore is rarely seen indoors.

Wet Rot often occurs in window sills or the base of exterior door

Dry rot. Buckled skirtings and cracking paintwork are often the first signs householders have that rot is attacking their property. The top side of these floorboards does not look too rotten, but when they were lifted, a severe attack by the dry rot fungus, *Serpula lacrymans*, was discovered underneath.

frames, where the paint film has cracked and rain has entered.

Typically water penetrates hairline cracks in paint, cannot evaporate again because of the film so the moisture content builds up in the wood to the point at which Wet Rot can start.

Control Measures

Dry Rot can spread throughout a house but Wet Rot although more common, usually confines itself to the originally damp site. All timber affected by Dry Rot must be cut out and burned and it is wise to ask a specialist to trace the full extent and deal with an attack if it is at all extensive. Free surveys are available for this and expert technicians will

Poor property maintenance is often the cause of timber rot. Here, water pipes under a floor burst, soaking timber joists and floorboards which are seen covered in the white, cotton wool-like mycelium of the dry rot fungus, *Serpula lacrymans.*

replace and treat with fungicide any affected timber. It is now also possible to insure against the cost of damage caused by fungal decay. Much business comes to professional dry rot contractors from 'botched up' attempts at amateur treatment.

Correct diagnosis is essential and it must be remembered that if the damp timbers suitable for Wet Rot are only partly dried out they may become vulnerable to attack by Dry Rot.

The source of any dampness causing rot must, of course, be put right and a frequent cause of conditions leading to Dry Rot in ground floors is rising damp in walls, due to a defective damp proof course. Three well-proved techniques are now available which can cure this without much inconvenience. The electro-osmotic system short circuits the small but measurable electrical difference between a damp wall and the ground on which it stands. This prevents capillary rise of moisture through the porous brick or stone, and the wall soon dries out.

A patented system can introduce an impermeable polythene membrane into the affected wall and a silicone injection system produces a waterproof barrier in brickwork. Many British local authorities also recognise the systems as eligible for an Improvement Grant in old property. Condensation can also contribute to damp conditions and attention may have to be given to heating, ventilation and correct insulation.

More obvious sources of dampness are missing roof tiles, broken guttering or downpipes, leaking plumbing or sanitary fitments or blocked airbricks. Make sure the original damp proof course is not bridged by a flower bed built up over it or a new path raised above it.

Special Dry Rot Fluid is available for the treatment of timber around outbreaks of dry or wet rot and all replacement timber. Use it liberally, but in the case of dry rot, first cut away all wood for 60 cm (2 ft) beyond the last visible sign of attack and remove plaster for 1 m (3 ft) beyond the last trace of fungal hyphae.

The fruiting body or sporophore of the dry rot fungus, *Serpula lacrymans* is very distinctive with its white edge and rust red centre. It is this part of the fungus which produces the spores that allow the fungus to flourish.

The whole area of attack should be opened up, brushed down with a wire brush and the decayed material removed from the building by the shortest possible route. All affected timber should be burned. Rubble can be sprayed with a fungicidal fluid as an extra precaution.

The correct disposal of the debris of decayed material by spraying with fungicide and burning is important because only by doing so can you ensure that further infection is not spread to other areas or other buildings.

If a wall is affected by the dry rot strands (or hyphae), a series of 12 mm (½ in) holes sloping downwards at an angle of 45° for about 15 cm (6 in) at 60 cm (2 ft), staggered centres should be drilled into it with a power drill so that the area can be irrigated with the fluid. The surfaces and all timbers within 1.9 m (5 ft) of the cut-out portions need to be soaked with two coats of a reliable fungicidal fluid. If timber is to be replaced in contact with the wall, the wall surface may be given a coat of zinc oxychloride paint or plaster as an additional fungicidal barrier.

In applying the fungicide, work from the highest level, downwards, spraying all brick, block, concrete and earth surfaces until they are saturated. A coarse spray is best for this job, either from a knapsack sprayer or a hydraulic pump. Where hyphae have penetrated brickwork both sides of the wall must be treated and it is sound practice to irrigate the wall as described, filling the holes with the fluid by means of a funnel. A 'toxic box' of holes around the perimeter of an affected area is often sufficient in conjunction with surface spraying.

Treatment of dry rot is not a job for the amateur. Get specialist advice from an experienced reputable company who have the resources to back a long term guarantee on their work.

All replacement timber must be thoroughly treated with a reliable fungicidal wood preservative, sawn ends being steeped in it, before they are placed in position.

The wall must be allowed to dry out thoroughly before redecorating. Apply one or two liberal coats of proprietary Dry Rot Fluid to all timber surfaces adjacent to the area of cutting away, to a distance of 1·5 m (5 ft) from the furthest extent of the cut away timber.

Protect any joists ends with a fungicide and a coat of bituminous paint before re-setting them into walls.

Rendered surfaces should be composed of a cement, lime and sand (1:1:6) mixture. As an additional precaution, it is a good plan to apply 6 mm ($\frac{1}{4}$ in) coat of zinc oxychloride plaster over the rendering coat (where a setting coat of wall plaster is subsequently to be applied) to an area extending 30 cm (12 in) beyond any previously attacked timber.

If necessary, fit new air bricks, or ventilators or clear any that have become blocked to ensure adequate ventilation. The metal grille type are the most effective for passing air through.

The treatment of Wet Rot is less drastic than that required for dry rot, and as long as the cause of dampness is removed and the timber allowed to dry out, no further growth of the fungus will take place. Test all timbers in the area of fungal attack with a strong, pointed instrument to determine the extent of sub-surface breakdown. Cut out and burn all timber which has suffered surface or subsurface breakdown due to fungal attack, together with all dust, dirt and debris.

Always select thoroughly dry, well-seasoned timber for replacement, cut it to size and give it two liberal brush- or spray-coats of a good Dry Rot Fluid or clear wood preservative over all the surfaces. Also, coat the adjacent surfaces of existing timbers and brick, block and concrete areas before placing the replacement timbers in position and fixing.

A good coating of a proprietary fungicide and immersion of the cut end-grains provides effective protection. This pre-treatment of new timber is a vitally important part of any remedial work and must be carried out after the timbers have been cut to size. Cut end-grains should be immersed for five minutes in the preservative to ensure good penetration.

A recently developed joinery injection technique avoids costly replacement of windows, by arresting wet rot in window frames and protecting against further decay.

Care of Outdoor Woodwork

In the garden, sheds, gates, seats and rustic work will all last longer if protected from decay and for such items an organic solvent type of wood preservative should be painted on in two liberal coats or sprayed on. If dipping is convenient, small dimension timbers should be dipped for 3 to 5 minutes and large timbers for 10 minutes. Wooden post ends to be sunk into the ground should be steeped for at least an hour or better still, overnight.

Organic solvent wood preservatives are available in clear, brown, or green, the green being specially made for wood that may come into contact with plants.

Water is a major enemy of outdoor woodwork. Exterior joinery, cladding and fencing swells as it absorbs water, shrinks as it dries and this sets up stresses that break joints open and allow fungal spores to get into contact with the damp wood. The mould *Pullularia pullulans* causes dark stains on wood and will even break out through varnish. It is now possible

Life cycle of dry rot fungus, *Serpula lacrymans*.

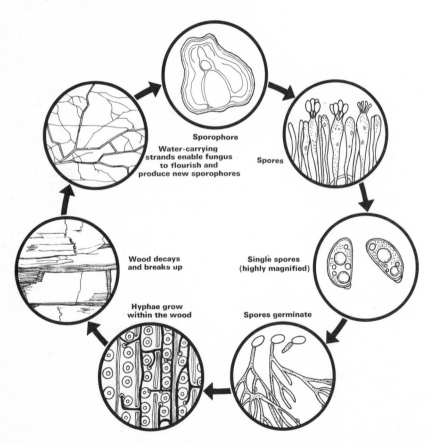

Sporophore
Water-carrying strands enable fungus to flourish and produce new sporophores

Spores

Wood decays and breaks up

Single spores (highly magnified)

Hyphae grow within the wood

Spores germinate

A water-repellent finish
(*left*) compared with
untreated wood (*right*).

to buy a cedarwood water repellent finish which is ideal for maintaining
the rich colour of Western Red Cedar—or for giving cedar colour to
other outdoor timbers. This stabilises the wood and prevents cracking as
well as discolouration by weathering. It stops wood going grey, preserves
it, and cannot peel, flake, crack or blister.

Mould, Condensation and Insulation

It is surprising what you can grow in a house with condensation
problems, without really trying.

It is curious fact that as standards of amenity and comfort in our homes
rise, so do complaints about mould growth on plaster, paint and
wallpaper in houses.

Mould is a general term used for a group of fungi that grow on the
surface of a wide range of materials. The commonest are species of
Aspergillus, Penicillium, Paecilomyces, Alternaria or *Cladosporium*, but
unless you are a mycologist you know them simply as green, black or pink
moulds.

The air around us is full of microscopic fungal spores which will
germinate and grow if they land on suitable surfaces that can provide the
food and moisture that they need. When they mature the moulds produce

Mould on plaster.

millions of coloured spores which are released into the air to start the process all over again. These lowly fungi belong to the division of the vegetable kingdom called Cryptogams, for which the Greeks had two words meaning 'hidden marriage'. The poor things reproduce asexually, but nonetheless effectively.

Very little food is needed by a mould but the food must have a carbon content. Paint, protein paint thickeners, cellulose or simply dust on the walls will supply sufficient so it is not practicable to starve the little moulds to death. It is, however, possible to eliminate mould by depriving it of water. First wipe down affected surfaces with antiseptic or a fungicidal solution to remove existing growth.

If we can eliminate water from the surfaces of our homes, mould spores will not germinate and we can prevent further trouble.

Moisture on or within walls may come from rising damp in old property, moisture of construction in new homes not fully dried out, faulty plumbing or exterior drainage systems, or most probably, from condensation.

Today, the cure of rising damp in walls is far easier than it was 25 years ago. Insertion of damp proof membrane by a silicone injection technique which forms a chemical barrier, or installation of an electro-osmotic

A severe case of rising damp causing disfiguring mould growth in a 12th-century Sussex church.

damp proofing system by a contractor who gives a proper 30 year guarantee, involves little of the mess and inconvenience of older traditional methods.

In a new house, mouldicidal paints, plenty of ventilation and adequate heating should cure the problem as the water of construction evaporates, although local damp patches such as those caused by a cold concrete lintel over a window may need further attention.

If outside gutters, downpipes and brickwork are kept in good condition and the gutters cleaned out once a year, no water should penetrate the house. Very exposed walls can be protected by rendering or the application of a suitable clear water repellent, which must however allow water vapour to pass out through it.

Condensation, however, is a different type of problem.

Basically, the problem depends upon the thermal efficiency of the building construction, the living pattern of the occupants and the effectiveness of the heating and ventilation. Many a draughty old house

143

Inserting a 'Discovac'
damp-proofing system in
a brick wall.

Silicone injection provides a chemical damp-proof barrier in brick walls.

Mould growth on ceiling
caused by condensation.

that was free from condensation has become a mould farm after a slick
conversion blocked up the fireplaces, sealed all the cracks and lowered the
ceilings.

We live in a humid climate and dry air in our homes would be very
uncomfortable so we need to live with atmospheric moisture but we can
do without the streaming windows, damp walls, mould on plaster, window
frames and furnishings that are some of the signs of the condensation that
occurs all too often.

Without proper understanding of what causes condensation complaints of creeping black mould are likely to continue.

Condensation is water that has literally appeared out of the air. All air
contains some water as vapour. Air being a mixture of gases, expands in
volume when heated, contracts in volume when cooled. Warm air holds
more water vapour than cold air. When warm moist air is cooled, or
touches a cold surface, the invisible water vapour condenses as visible
droplets of water.

Dew, or condensation, forms when the atmosphere's temperature
drops so that it can no longer retain the water as vapour and is therefore
saturated. The temperature at which this occurs is known as the Dew
Point.

Relative Humidity is the measurement of water vapour in the atmosphere. It is expressed as a percentage of the Dew Point which represents 100 percent relative humidity (RH). Mould tends to grow at anything over 74 per cent RH.

Get into your car on a cold morning and your breath mists up the windscreen. Only the heater and blower, or opening a window will disperse it.

Victorian houses with high ceilings, wide chimneys and howling draughts may have been uncomfortable but rarely had condensation. Fuel was cheap, so there was plenty of warmth and ventilation.

Modern houses, apartments and flats are efficiently sealed against draughts, any heat comes from radiators needing no fireplaces, —and they are often left unoccupied and unheated for most of the day, with all the windows closed. As a consequence, the structure of the little boxes in which many of us live never really warms up. Walls and ceilings remain permanently cold surfaces, just waiting to condense the water vapour poured into the air when we start boiling kettles, cooking, washing up and bathing and drying clothes, all within a few hours.

Water vapour behaves like any other gas and seeks to equalise its pressure by moving from an area of high pressure moist air to a low pressure dry area. In other words, steamy air from the bath or kitchen will drift into other parts of the home and may pass through porous materials such as brick and plaster, including ceilings.

Extractor fans can help to reduce local clouds of water vapour but a hot bath that will saturate 3,000 cu. ft (84 cu. m) of air at 19°C (68°F) will need a fan capacity of about 2,000 cu. ft (56 cu. m) per hour to reduce the relative humidity to 60 percent.

In an uninsulated house only 25 percent of the expenditure on space heating actually heats the rooms. The rest escapes through the walls, roof, doors and windows. Loss through walls and roof totals about 45 percent.

It is significant that people who have had the roof and cavity walls of their home insulated with mineral wool to save fuel costs have found that their condensation problems have also diminished or disappeared.

The insulation material, consisting of tiny fibres of mineral wool produced to a precise specification, is blown into cavity walls where it traps tiny pockets of still air just as a bird does when it fluffs out its feathers against the cold.

Cavity wall insulation with dry mineral wool.

A few cores are drilled out of the exterior wall and replaced after the wall has been filled, carefully matched so as to be virtually invisible. The effect is that of putting a tea cosy round the house, and with roof and walls insulated, fuel costs with thermostatically controlled heating can be reduced by a third. It is, therefore, well worthwhile getting an estimate for insulating your home whether you are growing mould or not. As fuel costs increase, the value of correct roof and wall insulation becomes more apparent.

The mineral wool, being a dry material, cannot shrink or crack within the wall cavity and is guaranteed not to transmit water through the wall. Roofs should have 4 in (100 mm) of insulating material between the joists either as loose fill or fibreglass blankets.

By keeping the indoor structural surfaces at a more even temperature and preventing wasteful heat loss, insulation of this type will very often reduce or eliminate condensation. Every cavity filling installation must be notified to the local authority in the UK and the contractor should have an official Agrément Board Certificate or, in the case of plastic foam, meet British Standard 5618.

If a house has solid walls without cavities, it is still possible to get some relief by insulating the roof and dry-lining the walls of rooms. Double glazing will provide extra comfort and reduce the condensation problem on windows, but is expensive to be considered solely as a means of saving fuel.

It is usually better—and no more expensive—to run the central heating on a relatively low thermostat setting all day than to run it flat out in bursts morning and evening. Once the house structure is warm, it will need little heating to maintain an even temperature, especially if the roof and walls are retaining the warmth indoors, where it belongs.

Under the Homes Insulation Act 1978, roof insulation with approved materials may be eligible for a grant of 66 percent of the cost up to a maximum of £50 in the UK.

Casual Intruders

Cricket

The House Cricket *Acheta domesticus* is about 12 mm ($\frac{1}{2}$ in) long, grey brown and shaped like a grasshopper with the two hind legs adapted to jumping.

It seeks warmth indoors in Winter in Europe or lives and breeds in decomposing refuse which is producing heat.

The characteristic 'chirping' by the male can be irritating in large numbers and it is sometimes necessary to use an insecticidal spray or fog on refuse tips or in heating ducts of large buildings. It is otherwise harmless.

Devil's Coach Horse

This long, black predatory beetle sometimes comes indoors hunting smaller insects. It is up to 3 cm (just over 1 in) long, with short wing cases which leave the abdominal segments exposed and when threatened the beetle curves its tail upwards like a scorpion, in a threatening posture. It has no sting, however, and its bite is not powerful enough to pierce human skin. It is harmless indoors, so put it back in the garden.

Earwig

Earwigs are often brought into homes on cut flowers and are frequent intruders from the garden, usually in search of suitable crevices as hiding

Earwigs with wings
extended.

places. They normally emerge only at night and cause no real damage in
houses.

Identification

The Common European Earwig *Forficula auricularia* has a long brown
body 10–14 mm (about $\frac{1}{2}$ in) in length with the characteristic 'pincers'
adding another 4–8 mm ($\frac{1}{8}-\frac{1}{4}$ in). It has wings but rarely flies. The
function of the cerci or 'pincers' at the end of the abdomen seems to be to
warn off predators but can be used to hold insect prey. They are straight
in the female earwig, curved in the male.

Life Cycle

The female lays a batch of 30 eggs in cells in the earth. The female stays
with the eggs in the 'nest' until they hatch. There is no larval stage, the
nymphs growing by a series of four moults. The adult sometimes
hibernates through the Winter, but will stay active in central heated
premises.

Damage Caused

A harmless wanderer from the garden where it feeds on other insects and plant tissue and may bite holes in flowers, but may be psychologically disturbing because of its alleged liability to make its way into the human ear. Dr N. E. Hickin states that reports of this are known to him, although the circumstances must be very exceptional.

Control Measures

Dust lindane or carbaryl insect powder in the humid areas where earwigs are found and remove creepers and herbage from around walls of the house. A spray of malathion around the outside of doorsteps and window frames and nearby vegetation may reduce the likelihood of infestation. Check pot plants and flowers brought indoors.

Ground Beetles

Various large black or violet beetles that occasionally wander in from the garden or emerge from under door mats. The larger ones grow up to 25 mm (1 in) long.

Remedy. They're harmless and no treatment is necessary but insecticidal dust may be used around points of entry into buildings.

Churchyard Beetle

This large, black, slow-moving beetle *Blaps mucronata* about 22 mm (nearly 1 in) long occurs in cellars, stables and other damp locations, moving about at night.

The larvae feed on vegetable matter. A harmless intruder.

Hoverfly

The commonest Hoverfly is probably *Syrphus ribesii* and because of its

Hover Fly.

yellow and black colouring is sometimes mistaken for a Wasp.

The characteristic hovering flight, with occasional abrupt slide-slips is typical, but it can also fly straight and fast.

Hoverfly larvae feed voraciously on aphids, so this insect should be regarded as a friend.

Lacewing

The pale green Lacewing or Lace Clearwing *Chrysops carnea* is a harmless wanderer from the garden or woods where its larvae prey upon other insects.

It has a pale green soft body about 15 mm (over $\frac{1}{2}$ in) long with richly

Pigeon fouling on factory roof girders.

veined transparent wings folded over it when at rest. The eyes are an
iridescent bronze. The adults are attracted to light and may enter houses
in autumn seeking hibernation sites.

No treatment is normally necessary.

Pigeons

The feral or 'London' Pigeon is a classic example of the human race
creating its own pests. A bird of very mixed parentage it is descended
from the cliff-nesting Rock Dove *Columbia livia* via the old manorial
dovecotes of England—and after all the next best thing to a cliff in
modern towns and cities is an office block, public building, railway arch
or apartment block.

153

Identification

Apart from the larger Wood Pigeon, which has a conspicuous white half collar and the smaller Collared Dove, which has a black half collar, any pigeon encountered around buildings may be assumed to be a feral bird. The majority are grey and black with a purple or green sheen but inter-bred 'red chequered' pouters, fantails and other fancy varieties have produced a range from pure black to pure white.

Life Cycle

The normal brood is of two white eggs, laid at almost any time of the year in a rudimentary nest of twigs, straw, bits of string and any convenient urban debris.

The nestlings or 'squabs' are helpless after hatching but are fed by regurgitated food — 'pigeons milk' — by the parent birds.

Damage Caused

With few natural predators, the diseased pigeons survive as well as the fit ones and can transmit ornithosis, histoplasmosis and other diseases to humans. Accumulated droppings, dead birds and nest debris foul buildings, cause damage to guttering, drainage and stonework and provide breeding sites for flies, mites, carpet beetles, and many other insects. The droppings also make areas under perches hazardous and slippery.

Control Measures

A soft plastic bird repellent jelly will repel pigeons and starlings if applied correctly where the birds alight. It does not harm the birds or buildings.

Bird control contractors may use this repellent or large nylon nets to protect a building. They are also empowered to destroy pest pigeons after laying a stupefying bait or by using cage traps.

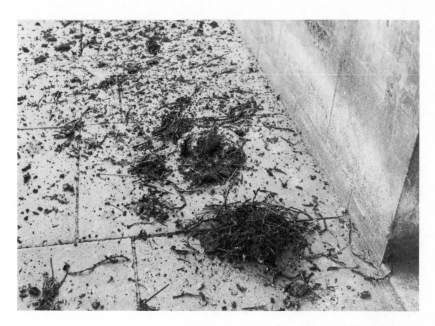

Nest debris from pigeons that blocked roof gutters on a London building.

Applying soft plastic
pigeon repellent.

Scorpion.

Scorpions

There are no British or North European Scorpions and *Euscorpius italiens* is one of the few species in Southern Europe, none of which is dangerous but their sting has been compared to that of the Hornet.

Some tropical species are more venomous and medical advice should be sought if a person is stung.

Scorpions feed on insects and spiders and are normally active at night, remaining concealed under rocks or in crevices in houses during the day. The females produce live young.

Being large creatures, they require a heavy dose of insecticide if chemical warfare is preferred to physical assault. They are really best left to the experts—and if scorpions are around, shoes or boots should be shaken out before putting a foot in them.

Related Species

Tiny 'Fake Scorpions' prey on small insects in neglected parts of buildings, but are too small to bite through human skin.

Woodlice

Harmless grey segmented creatures, they are not insects but crustaceans. One species rolls up into a characteristic tight ball when disturbed. Usually associated with cool damp conditions in corners of bathrooms, kitchens or under doormats.

Identification

Woodlice have oval, grey bodies of regular segments. They have fourteen legs and prominent antennae. They are also known as Pill Bugs, Slaters or Sow Bugs.

Three species are very common in the garden and may enter houses. *Oniscus asellus* is the largest, about 15 mm ($\frac{5}{8}$ in) long and has smooth segments, *Porcellio scaber* is narrower with dull, rough segments, and *Armadillidium vulgare* is the 'pill wood louse' which rolls up into a tight ball when disturbed.

Damage Caused

Woodlice, the only really terrestrial crustaceans, need moist, cool, dark places to live. They are harmless indoors and in the garden they feed on very rotten wood or other vegetable matter.

Control Measures

Individuals are susceptible to any good household insecticidal powder or spray and remedying the cause of dampness will prevent recurrence. It may be necessary to cut vegetation back from around the outside of the house.

PART III   The Broader Viewpoint

Insects in Perspective

We curse the horse-fly that lands on our neck, we spend vast sums of money to eradicate locusts, mosquitoes, tsetse flies, termites and other insect pests, yet without the teeming millions of insects, the world that we know could not exist.

In the forests of the Carboniferous age, 250 million years ago, flying insects were already winging their way between the giant ferns and horsetails that now form our coal deposits. The only invertebrates to develop the power of flight, when insects grew wings they made a great stride in evolution that put them ahead of all land Arthropods.

Most of the recognisable types of insects had developed by the time the first mammals appeared 150 million years ago and by the time insect bodies were becoming trapped in amber, 20 million years ago, almost all families had evolved, except those that are parasitic on humans who only turned up about a million years ago.

Even today the most primitive Orders of insect, the springtails, bristletails and silverfish have no wings. Some of the higher Orders, such as crickets and earwigs, only use wings occasionally, some such as bed bugs have only vestigial wings and some, such as lice and fleas, have lost the wings which would have been an encumbrance to them.

Insects now outnumber all other forms of animal life both in total numbers and in diversity of species. They include the good, the bad and the ugly. On the face of it there is little in common between the delicate grace and beauty of the Swallowtail butterfly and the squat ugliness of the Crab Louse, yet both are equally insects.

One of the factors leading to this diversity within the Class Insecta is the short life cycle of insects. Taking an average life of six months, compared with five years for vertebrates, there will be ten times as many generations of insects in a given time and these will be subject to correspondingly more mutations. This fact, together with their wings, gave insects a truly 'flying start' in the evolutionary race.

A million species of insects have been identified and there may be several times that number to be discovered and classified. This compares with about 10,000 known mammals and about 15,000 birds. The largest group of insects, the beetles has half a million known species. There are 100,000 butterflies and moths. The structure and physiology of insects enables them to achieve remarkable feats of athletics and endurance such as carrying loads many times their own weight and jumping distances many times their own body length. Special long-chain proteins produce the elastic substance responsible for the extraordinary jump of the flea. Other extraordinary chemicals called pheromones govern the behaviour of some social insects. The female Gypsy Moth produces a sex attractant odour that can be detected by male moths over tremendous distances. The male Lobster Cockroach produces a sex attractant named Seducin by entomologists. This is presumably the cockroach equivalent of after-shave lotion.

The adaptability and specialisation of insects has enabled them to colonise the Earth from the freezing Arctic to the burning desert, although the largest and most spectacular forms are in some of the tropical rain forests. Here, the South American beetle *Titanus gigantus* attains a length of 13 cm (5 in) or more.

There are insects that feed on materials that would kill a person. There are insects that bore through sheet lead, others that feed on dead museum specimens, and a fly maggot that can only survive in crude petroleum.

There are parasitic wasps that detect the body of a host grub deep inside a piece of wood and there are butterflies that depend on colonies of ants to rear their caterpillars.

Fleas and lice have altered the course of history, spreading The Black Death and the scourge of Typhus. Locusts, Tsetse Flies and Mosquitoes still make large areas of the world uninhabitable to us. Others have proved our ally in destroying or controlling pests or weeds.

The Kalahari Bushmen make deadly poison for their arrows from the pupae of the South African leaf beetle *Diamphidia locusta*. Man cannot produce a 'cold' light, yet Glow worms and Fireflies have been lit up by their own chemicals for aeons of time.

Chemical warfare is practised by the Bombardier Beetle and by the nasutes of certain termites. Other termites build their own cities with controlled temperature and humidity and have specialised castes to carry out the various duties needed by the community.

Dr Norman Hickin quotes a report that lightly roasted termites bought in a Congo market had a higher calorific value as food than steak or cheddar cheese.

The Scarab was sacred to the Ancient Egyptians for whom it symbolised life itself. Perhaps they knew a thing or two, for although we may admire the symmetry of a honeycomb or mourn the disappearance of some butterfly that coloured our schooldays, we tend to overlook the importance and to underestimate the significance of insects and their relationship to all other forms that make up the intricate web of life. The proper field of study of the human race may prove to be ecology.

Efforts over the last few years have begun to create a public awareness of the complex food chains upon which our own lives depend and within which insects form a number of essential links.

Other less publicised aspects of insect activity include their role as pollinators. Without insects, fruit would not set and all insect pollinated plants would die out. Even the colour, shapes and scent of most flowers have been evolved to attract insects to pollinate them.

By contrast, insects also bury the dead, scavenge and destroy decaying matter as Nature's refuse disposal teams. The burying beetles, flies and other species quickly and efficiently render bodies to dust and valuable humus aided by bacteria, fungi and worms.

The published literature on insects is vast, dating back to Aristotle and Pliny. Men like Henri Fabre have devoted their lives to study insect behaviour. Dr Thomas Mouffet (or Muffet) who wrote the Theatrum Insectorum in the 17th century is little known except by entomologists, but his daughter Patience is immortalised as 'Little Miss Muffet' of the nursery rhyme.

Sources of Further Information

National Poisons Information Service

This service is provided *for medical practitioners* who can obtain advice on toxicity, symptoms and treatment of pesticide poisoning if it should occur.
The service operates from the centres listed below.

	Tel. No.
England	
National Poisons Information Centre,	
New Cross Hospital,	
Avonley Road,	
London S.E.14.	01-407-7600
Newcastle Poisons Information Bureau,	
Royal Victoria Infirmary,	
Queen Victoria Road,	
Newcastle 1.	0632-25131
Leeds Poisons Information Bureau,	
General Infirmary,	
Great George Street,	
Leeds 1.	0532-30715
Scotland	
Scottish Poisons Information Bureau,	
Royal Infirmary,	031-229-2477
Edinburgh.	(Ward 3)

Wales
Cardiff Poisons Information Centre,
c/o Medical Records,   0222-492233
Cardiff Royal Infirmary,   (Poisons
Cardiff, Glamorgan.   Enquiry)

Northern Ireland
Belfast Poisons Information Centre,
Casualty Department,   0232-30503
Royal Victoria Hospital,   (Poisons
Belfast.   Enquiry)

Eire
Dublin Poisons Information Centre,
Jervis Street Hospital   Dublin
Dublin.   759 197 45588

Some Further Information Sources

The British Pest Control Association,
   Alembic House, 93 Albert Embankment,
   London SE1.

The British Wood Preserving Association,
   Premier House, 150 Southampton Row, London WC1B 5AL.

Ministry of Agriculture, Fisheries & Food,
   Pest Infestation Control Laboratory,
   Government Buildings,
   Hook Rise South,
   Tolworth, Surbiton, Surrey, KT6 7NK.

The Royal Institute of Public Health and Hygiene,
   28 Portland Place, London, W1.

The Health Education Council,
   Middlesex House, Ealing Road,
   Wembley, Middx., HA0 1HH.

The Environmental Health Officers Association,
  19 Grosvenor Place,
  London SW1X 7HU.

The Hotel, Catering and Institutional Management Association,
  191 Trinity Road,
  London SW17 7HN.

The Rentokil Advice Centre,
  Felcourt,
  East Grinstead, West Sussex RH19 2JY.

Identification Service

Confused Flour Beetles, Wharf Borers, Varied Carpet Beetles, Yellow Swarming Flies, Golden Spider Beetles, Silverfish and even Bed Bugs are among the items that arrive in the post at one large pest control company's laboratories. Here entomologists examine match-boxes, jars or tubes containing insects sent in for identification. Over 22,000 different insect species live in the British Isles and each year over 1,000 requests for information are received, accompanied by insects, mites or other animals.

The laboratory also receives fungi, samples of infested timber, pieces of wood for chemical analysis and sometimes just dirt from machinery and floor sweepings. Not infrequently the insect is somewhere in a packet of potato crisps, or in a piece of cake or bag of nuts, having got into the food at some stage before or after manufacture. How and when the insect got there is a frequent question which often requires some astute detection. The answer may well influence a Court decision. Recently the bone of a Western African snake was found in a bag of 'Pure Californian Peanuts'!

Considerable knowledge of the terms used to describe various parts of an insect is required to enable the proper use of identification keys. Are the 'prosternal carinae present or absent', or the 'metasternal epimera entirely yellow?'

About half the insects received are beetles, about 10 percent flies, and the same number are moth grubs, which if alive, have to be killed in alcohol to show the hairs—some less than a thousandth of an inch long (0.025 mm)—which distinguishes even the more usual species. Dead, dried up specimens are more difficult to name and these have to be warmed in a special solution which swells them to their original shape. All adult moths which attack stored foods and textiles tend to look similar after their scales have been brushed off during the process of amateur catching and killing.

Naming mites really is a job for the specialist. As the name suggests, these creatures are extremely small, usually only 0·5 mm ($\frac{1}{50}$ in) long. They have to be warmed for a few minutes in acid and then examined under a high-power microscope which magnifies them about 500 times.

Whether or not a customer receives an identification of a specimen sent to the Laboratory depends on its 'shape' on arrival. Insects reach the entomologists in all conditions; they may be H.O.F. (hammered out flat), S.O.S. (stuck on Sellotape), mixed with pounds of wheat, or missing altogether.

The laboratory's constant plea is for people not to put woodworm in matchboxes. The matchbox is usually empty on arrival, with a tell-tale hole in one side!

At least one original contribution to science has been made by the identification service which a few years ago described an entirely new species of carpet beetle sent from the Channel Islands. Several specimens of it have since been reported from London and Southern England.

More recently the Laboratory was able to reassure a lady that the small brown objects 'infesting' her bed were the broken plastic tips of her hair grips.

The Pied Piper — A Legend Explained

'Hamelin Town's in Brunswick
By Famous Hanover City;
The River Weser, deep and wide,
Washes its wall on the Southern side'.   Robert Browning

The legend of the Pied Piper is familiar throughout the English speaking world and was part of German folklore long before Browning wrote his famous poem.

Only recently, however, has a satisfactory theory been introduced that explains the origins of the story and at the same time makes it clear that the character whose 'queer long coat from head to heel, half of yellow and half of red' is Death.

The tale we heard at our mother's knee was of a town over-run by rats that were piped away and drowned in the river by the Pied Piper. When he was refused his fee he led all the children away into a cave in the Koppelberg Hill which closed after the children and they were never seen again.

Past scholars have suggested that the story describes the Children's Crusade of 1212, when one Nicholas from Cologne travelled through Saxony and led a party of 20,000 young people away to try and reclaim the Holy Land for Pope Innocent III. They apparently disappeared in Italy. However, the latest theory is that the legend is an account of a visitation by The Plague.

Professor D. Wolfers formerly of the London School of Hygiene and Tropical Medicine points out that there were two outbreaks of Plague in Hamelin and Hanover, thirteen years apart, in 1348 and 1361. Carried by rat fleas, the disease would kill a great number of the rats and what more natural than for the citizens to sweep the ravaged corpses into baskets and tip them over the south wall into the swift-flowing Weser? Hamelin, a milling town, had more than its share of rats in those days.

Bubonic Plague confers a good degree of immunity upon those who survive an attack and the survivors of the 1348 epidemic were probably little affected by that of 1361. This outbreak would, however, have taken its toll among the children under thirteen, possibly killing all save one who according to Browning's poem was lame and could not join his playmates. The legend gives the death roll as 130.

'When lo, as they reached the mountainside, a wondrous portal opened wide,
As if a cavern was suddenly hollowed;
And the Piper advanced and the children followed'.

With death taking so many, so quickly, it would be natural as well as prudent to arrange a mass burial of the unfortunate children. Could not this be the hole in the Koppelberg Hill of the legend? Such a dismal cemetery would be long remembered and as for the representation of the Piper, it was common in Medieval art to depict Death as a dancing skeleton, leading as string of people to their final fate.

D'Israeli writing in 1840 states: *'The Gothic taste of the German artists, who can only copy their own homely nature, delighted to give human passions to the hideous physiognomy of a noseless skull; to put an eye of mockery or malignity into its hollow socket, and to stretch out the gaunt anatomy into the postures of a Hogarth; and that the ludicrous might be carried to its extreme, this imaginary being, taken from the bone-house, was viewed in the act of dancing'.*

As this morbid art-form developed it became stylish to depict the dancing skeleton playing a musical instrument and the coloured dress of the Piper may well be accounted for by the buboes, haemorrhagic patches and pallor characteristic of victims of Bubonic Plague.

Although Browning gives the date of the Hamelin event as July 22nd 1376, this appears to be an estimate derived from the first coherent written description of the story by Dr Fincelius who wrote in 1556: *'of Devil's power and malignity I will tell you a tale. About 180 years ago in Hammel, on the Weser in Saxony, on St. Marguerite's Day ......'.*

We may thus be fairly sure that it was not exactly '180 years ago' and indeed in Hamelin today the event is commemorated from a date of 1284. This apparently comes from a report by a traveller writing in the 16th century who described an archway in Hamelin with an inscription about events '250 years ago'.

There was a second smaller outbreak of 'Children's Plague', the *Pestis puerorum*, in Hamelin in 1375 which fits the first date better but does not seem to have been so dramatic as that of 1361.

Incidentally, the Pied Piper legend is not the only record of the Black Death in children's folklore. The 17th century nursery rhyme 'Ring a Ring O' Roses' dates from the Great Plague of London—'All fall down' graphically describing the effect of the disease.

The direct connection between rats and Plague, worked out in the 1890's was not of course understood in Medieval Saxony, yet might not some local alchemist have observed that the disease followed large rat infestations? He may even have told the Mayor and Corporation 'You've got to get rid of the rats!'

Perhaps he was unheeded and when events proved him right was begged to write out the Hamelin story that we know.

Perhaps indeed, suggests Dr Wolfers, he was unheeded because he had cornered the market in rat poison.

There is also an Oriental tradition that when Buddha died, the rat was charged with the task of carrying death to the rest of the world within 2,000 years.

The German Chronicler Verstegen's account of 1605 describes the legend almost exactly as we know it today, and in modern Hamelin it is re-enacted by the local children on an open air stage, after a procession through the streets. But the man who kept Hamelin free from rats for ten recent years was employed by the German subsidiary of a British pest control company.

The company, remembering the legend, had a contract which provided for payments quarterly—in advance.

And beside the Ratcatcher's house runs Bungelosen Strasse which means The Street Without Drums. No musical intruments have been permitted since, it is said, a Pied Piper's flute led the town's children down it for the last time over 600 years ago.

Bibliography

Insects and Hygiene (Busvine) Methuen, London.
Handbook of Pest Control (Mallis) MacNair-Dorland, New York.
Household Insect Pests (Hickin) Associated Business Programmes, London.
The Woodworm Problem (Hickin) Hutchinson, London.
The Dry Rot Problem (Hickin) Hutchinson, London.
The Cockroach (Cornwell) Hutchinson, London.
Pests of Stored Products (Munro) Hutchinson, London.
Hygiene in Food Manufacturing and Handling (Graham-Rack and Binsted) Food Trade Press, London.
Pest Control In Buildings (Cornwell) Hutchinson, London.
Problems of The Environment (Brooks) Harrap, London.
The Life That Lives On Man (Andrews) Faber, London.

Some Common Pest Control Chemicals

'The mosquito was heard to complain
That the chemists had poisoned his brain.
The cause of his sorrow
Was para-dichloro—
Dyphenyl—trichloro—ethane.   Dr R. A. E. Galley

These chemicals can be obtained under a wide range of proprietary brand names. When purchasing them it is important to check the main active ingredients, which will be listed on the packaging.

For Flying Insects

Diazinon
Dichlorvos
Pybuthrin
Pyrethrin

For Crawling Insects

Arprocarb
Bendiocarb
Bromophos
Carbaryl
Chlordane
Dieldrin
Fenitrothion
Iodofenphos
Lindane (Gamma HCH)
Malathion
Pirimiphos methyl
Pyrethroids (permethrin)

For Wood Boring Insects

Dieldrin
Lindane (Gamma HCH)

For Textile Pests

Dieldrin
Lindane (Gamma HCH)
Paradichlorobergene

Fumigants

Aluminium phosphide
Methyl Bromide

For Rats and Mice

Alphachloralose
Brodifacoum
Bromadiolone
Calciferol
Coumatetralyl
Chlorophacinone
Diphacinone
Difenacoum
Norbormide
Warfarin

Index